THE WAY OF TACTICS

BY

TOM SOTIS

LIFE LESSONS AND THE MANIFESTO OF INVINCIBILITY FOR WARRIORS, EXECUTIVES, AND ATHLETES BY THIS GENERATION'S MOST PROLIFIC INSTRUCTOR OF KNIFE FIGHTING

ISBN 978-1-7348045-0-8

Copy editing, proofreading, formatting, and cover design by ebooklaunch.com
Published by Thomas Sotis

Thomas Sotis
6 Roger Street
Cumberland, RI
02864 USA
tom@tomsotis.com
www.tomsotis.com

Thank you for your interest in this book. If you would like to receive updates on future releases or want information, please sign up for my email notices by visiting http://www.tomsotis.com

To my wife Lucia and my son Jake,

who are my happiness and my joy.

Facing the Enemy

No design, no conception;
just reflexes and reactions
one with my blade,
alive in my hands
pure of heart and
firm in my stand
the moment you move
I know what to do
strike from the heart! and
I will strike true.

– *Tom Sotis*

CONTENTS

INTRODUCTION

The Way of Tactics is presented in three parts. "Life Lessons" is intended to entertain readers while also offering personal valuable insights, thereby giving context to the "Manifesto of Invincibility". The Life Lessons are not in exact chronological order, but are grouped in such a way as to improve continuity for readers.

A "Manifesto of Invincibility", presented through the prism of knife fighting, has useful applications for warriors, executives, athletes, and high-performance individuals. It presents a framework for achieving optimal human performance from within the roots of your deepest motivations which determine your strategies, guide your tactics, and dictate the course of your actions.

I state, unequivocally, that it *is* possible to be invincible. I do not believe in the figurative concept of sustained invincibility (where one person could defeat all others) but I am certain that the literal reality of momentary invincibility exists. This book presents the elements of invincibility and a formula for achieving it.

The "Applications" section presents my pioneering vision towards disrupting an industry, establishing a new classification of business, and raising it to professional standards. "Applications" goes on to explain how the Manifesto is a pragmatic tool for improved outcomes in business, sports, and life.

I hope every reader enjoys and benefits from this book.

Tom Sotis

PART I: LIFE LESSONS

LIFE LESSONS

The Flag

> "The American flag represents all of us and all the values we hold sacred. — *Adrian Cronauer*

I have only very vague memories of President Kennedy's assassination because I was only four years old. I wasn't old enough to understand what was going on, but I remember sensing that something terrible happened because the adults were really sad. I recall walking with my father and asking why the flag was only halfway up the flagpole. He explained that when a great man dies, people everywhere lower the flag as a sign of respect. The earliest memory burned into my psyche is that of the American flag, and I understood – at the level a four year-old can – the importance of a lesson that has only grown more meaningful with each passing year.

> *Life lesson: The American flag is a symbol of something greater than ourselves.*

Two Hands, Two Feet

> "You can be an underdog, but you can beat the champion."— *Joanna Jedrzejczyk*

In 1969, the Vietnam War was raging; the culture war was full-on with protesters, hippies, and peace signs, and in the Summer of Love, I was a bike-riding 10-year-old. Every night, images from Vietnam were on the news. My father always made favorable comments about our brave soldiers, and my mother always commented on "those poor people suffering so badly". They were both right, and I have carried those sentiments with me for life. Watching the older kids being drafted, all of us younger kids assumed that on our 18th birthday, we would be too.

When martial arts exploded into mainstream America with Bruce Lee, Billy Jack, and the Kung-Fu television series. In that era, the only protective gear we had was a mouth guard, groin cup, and thin foam shin pads. In the karate school, bloody noses and split lips were commonly expected occurrences in nearly every class.

And tournaments were no joke. Sure, there were points awarded, but contact to the face was permitted and you only got a warning if you drew blood. In a tournament, you fought strangers who were doing their best to hit you hard enough to put you out of the fight. It was not uncommon for full-on fighting to break out between the fighters, and on more than one occasion, brawls broke out between rival groups or schools.

One morning, on the day of a tournament, my grandmother asked me why I seemed annoyed about fighting in the tournament. I told her I was angry because it was unfair the way we (in junior division) were lined up to fight according to height with no regard for rank, and as a beginner, I would have to fight bigger and more advanced fighters above my skill level. Instead of coddling me with her usual grandmotherly hugs, she was stern when she imparted this:

Life Lesson: Whoever you fight, they have just two hands and two feet so you have the very same chance they do.

"And don't ever forget that. Now, go do your best." The gift of my grandmother's advice has since given me the confidence to prevail at times when I was at a marked disadvantage. In case you are wondering… I was outclassed and got my ass kicked that day, but I fought well, so winning or losing became, if not irrelevant, secondary to performing at my personal best, whatever the outcome.

Apples & TV

> "Humanity is an ocean; if a few drops of the ocean are dirty, the ocean does not become dirty." — *Mahatma Gandhi*

Growing up during the Cold War, I still have vivid memories of the elementary school alarm for a nuclear attack. We would huddle under our desks, covering the backs of our heads with our hands, and close our eyes, innocently believing we would be safe. I grew up learning that Russians were bad, and at any moment, they might obliterate *everything*.

Since early childhood I've had a lifelong love affair with *National Geographic* magazine. Even before I could read, I marveled at the photos from a wide variety of cultures and dreamed that someday, I would visit distant lands and meet interesting people. I have always held a special fascination for less technologically-advanced cultures and was curious about how inventive and creative humans adapted to the world around us. The world was so vast and life was so different everywhere. I wanted to go *everywhere*!

I remember one day telling my mother that I intended to see the world...except Russia, of course, because they were bad. That day, my mother taught me two important things. First:

> *Life Lesson: There will be a few bad apples no matter where you go, but mostly you will find good people doing their best to get through life the best way they know how.*

I told my Mom that on TV, they said Russia was bad. Then she gave me more advice:

> *Life lesson: You shouldn't believe a lot of what is said on TV.*

Always Show Up

> "Show up authentically and synergy will meet you there." — Sanjo Jendayi

In 1980, I was fighting in a tournament and scored an exciting point with a jump spinning kick just as Aaron Banks, owner of the

World Professional Karate Organization, happened to walk by. After the match, he introduced himself and asked who my coach was. It turned out that he already knew my coach well. They talked for a few minutes and then Aaron came back and said there was a title vacancy; on the spot, he offered me a shot to fight for the Light Middleweight Kickboxing Championship.

Plucked out of obscurity for a title shot was like being in my own *Rocky* movie! It was exciting, but the novelty wore off right away and I was left with the daily grind of grueling physical punishment and sparring with fresh partners in every round. In between training sessions, since I had to lose weight for the fight, my diet did not include cheeseburgers. For six months, I had actual dreams of biting into a juicy cheeseburger after the fight. I think perhaps the cheeseburger was a symbolic conclusion that helped me to resolve dealing with the pressure from the unknown outcome of the fight. Whether that is true or not is perhaps best answered by psychologists, but I can say that, win or lose, I looked forward to relishing an after-fight cheeseburger with the same joy of being rescued off a deserted island.

I didn't know if I deserved a shot at the title, but I was damn well going to be worthy of it. I knew my opponent was stronger than me, so my coach and I strategized to wear him down in the first half of the fight and increase the pressure in every round of the second half. I worked on increasing my long-term endurance. I pushed myself harder than I had ever done before. I ran every day, adding small increments of distance, until I was running for an hour, and then I began sprinting every few minutes within the hour. I worked out – not just to the point of exhaustion but until continuing required the kind of effort that was purely an exercise of willpower.

> "A real fight doesn't begin until the moment you want to quit but know that you must continue." — Tom *Sotis*

The morning of the fight, my opponent's coach called my coach to ask if we were willing to reduce the fight from ten rounds to five. When they asked me about it, I took it as a sign of weakness on my opponent's part and absolutely refused. I knew that meant he would give it his all, trying to knock me out in the first five rounds and not saving anything for the second half, so I was surely going to take even more of a beating in order to win in the second half. But that made zero difference to me because I knew his spirit was cracked before he even entered the ring, whereas mine was rock solid.

A few hours later, as we were getting ready to leave for the stadium, my coach got a call saying my opponent had backed out of the fight. The inconceivable had just happened! I couldn't imagine any fighter would train for six months and then back out, unless forced to do so by a medical issue. A fighter recovers from losing a fight, but he will never recover his honor after backing out. I was outraged from being deprived of the chance to fight that I had earned through all the years of training and effort. Win or lose, the fight was *mine*, and then, all of a sudden, it wasn't.

> "We must all suffer one of two things: the pain of discipline, or the pain of regret or disappointment." —*Jim Rohn*

Having the rug pulled out from under me was a major disappointment, but there was nothing I could do, so I gathered my things to go home. However, my coach said we had to go to the fight anyway. I asked, what the point of going if my opponent already forfeited? My coach replied that once I had committed to the fight, it mattered not what my opponent does, only what I do. That day, my coach taught me a lesson, and I didn't realize how important it was until much later in life.

> *Life lesson: Once you commit to a fight, you show up no matter what.*

The event was called the "Oriental World of Self-Defense" and featured a number of different demonstrations by masters of different martial arts. The promoter, Aaron Banks, was very happy that I showed up because my fight was the only fight of the night, and it was imperative that a fight took place. So, the promoter substituted in a fighter.

After the fight, I was interviewed by *World Karate Magazine;* I was asked what it felt like to be the champion. I said I didn't know because there were guys in my own neighborhood who could still beat me. They said that response wasn't fit to print, so I told them the interview was over.

Then, Aaron came over to take photos and congratulate me on winning the title. I told him I hadn't won because I had won by forfeit. In addition, the last-minute substitution had reduced the fight, in my opinion, to barely more than an exhibition. Aaron said he understood my disappointment because I was a fighter at heart, and he said that even though I didn't win in the ring, according to martial philosophy, I had won in the best possible way:

> "Supreme excellence consists of breaking the enemy's resistance without fighting. The greatest victory is that which requires no battle." — *Sun Tzu*

At the time, and for a long time after, Aaron's words were no consolation, though eventually, I have come to appreciate what he said, and with perspective, his words ultimately helped to assuage my bitterness.

After a long, painful journey to the ring, being disappointed by my opponent, fighting the substitute, and the interviews, it was almost over. Mentally and emotionally drained, I just wanted one thing to bring all of it to an end. I wanted closure ... I wanted a cheeseburger. After the show, we drove straight to a fast-food drive-thru for my long-awaited cheeseburger. Finally! While ordering our food, my coach's children kept changing their minds, confusing the girl taking our order and frustrating my coach to the point that he gave up and just drove off. No cheeseburger for me....

> "When life hands you a lemon, sometimes all you can do is suck on it." — *Tom Sotis*

Blocking Bullets

> "Belts don't stop bullets." — *Unknown*

The black belt fighting team from the karate school where I trained as a kid was undefeated in New England. Every member of the team was a proverbial "tough as nails" guy. Most of them had grown up in rough neighborhoods where fighting was part of living.

The team's unofficial captain was George "Bash" Basmajian. Bash was huge (strong-huge, not fat-huge). Lightning fast like you would never expect from someone his size, he displayed an obvious mean streak. Without a run or jump, from a *standing* position, Bash could execute a side-kick so hard it would split the heavy canvas bag and stuffing would pop out. They had to get a heavier leather bag with Bash around.

When I was a kid, Bash was my idol. Nobody messed with Bash and I wanted to be like him. Training in the school was always hard, but when the chief instructor retired and Bash took over, it became more like training to be a gladiator. It became increasingly brutal, where blood in class was not incidental, it was a *requirement*. On some nights, Bash would close the doors and nobody was permitted to leave until there was blood on the floor. On more than one occasion, someone would ask another to give them a bloody nose so we could get out. If my parents had known what was going on in that school, they would never have let me train there. But I stayed, as did my older brother John, and we watched the class attendance gradually decrease from about thirty guys to about six before Bash closed the school.

Soon thereafter, Professional Full Contact Karate was catching on. I was too young to fight, but the coach let me train in kickboxing, even though I couldn't get in the ring. In my first lesson, I asked the coach if I could spar with Bash. Coach warned me....

Bash asked if I was ready; I said yes, and he knocked me out cold with one punch. I came to with Coach shaking his head, saying he had warned me. I asked Bash if he'd had enough and wanted to take a break. He smiled and nodded. My head was pounding, but it felt great to earn his respect. George took me under his wing where I learned a lot about fighting, both in and out of the ring.

In Providence, RI during that era, organized crime was in its heyday with criminals and underworld figures enjoying celebrity-like status. Unfortunately, Bash fell in with the wrong crowd and became a hit man for organized crime. I was driving to school one day to take my sociology exam when I heard over the radio that Bash had been murdered the night before by other gangsters.

Life lesson: Nobody is tougher than a bullet.

If Bash was the scariest fighter I ever met, David Randall was the fastest. When we were teenagers, David was so fast that sometimes he would even tell me ahead of time how he was going to attack, and he'd still manage to hit me. Then, one day, I got a call telling me David had been shot to death outside of a nightclub over a gram of cocaine in a drug deal gone bad.

Life lesson reinforced: Nobody is faster than a bullet.

The Bottom Line

> "Anyone who thinks the pen is mightier than the sword has not been stabbed with both." — *Lemony Snicket, When Did You See Her Last?*

Once, when training retail store detectives in Los Angeles, a young man stole something. When he came out the door, I approached him, identified myself, and he bolted. At the time, I used to run ten miles almost daily so I was on his heels in a heartbeat; however, he was fast, and it was all I could do to keep up with him. I was hoping he'd run out of gas before me.

I chased him down alleys and through backyards, up the street, and then into a housing complex. As he ran through the courtyard, he started yelling for help. He ran past a few guys standing around who then started moving to block my path. As I got close to them I slowed down, raised my hand as though signaling I was going to stop, and then at the last moment, burst right through them.

Now I was chasing the kid and the guys were chasing me, but they quickly gave up. I'd had enough and doubled-down to catch this thief, pouring it on, and running with every bit of energy I had.

I was right on his heels as we leaped over a low wall. We landed in a courtyard and he was trapped. He drew a knife and I froze in my tracks. He said, "You touch me and I'll stab you." And so, I jumped back over that wall.

Life Lesson: When you face someone with a knife who is committed to using it, unless there is something you are willing to risk your life for, you will not engage them.

Red-Handed

"Having a guitar don't make you a musician." — *Unknown*

The first time I went to Mexico was in 1982. I was working undercover, attempting to locate the person at the end of a pipeline that began in Los Angeles. My partner "Ski" and I had the name, address, and old photo of a man we were supposed to find in Culiacán, 600 miles south of the border. We were to take a new photo of him.

My Fiat broke down in Tucson, Arizona, so I left it with a local mechanic and we obtained a rental. As we waited in line at the border in Nogales, we were approached by a man who asked if our car was a rental (the Rent-A-Car sticker gave us away) and he informed me that it was against the law to bring a rental into Mexico; you must own the car, and luckily, he knew someone that could help (for a fee of course). We followed him to a nearby store where the owner was said to have the official seal to press into ownership papers. I was wary, but it turned out OK. We got the papers, bought some acetone and razors, scraped the rental stickers off the car, and got back in line at the border. We passed through the border with no problems. A short way down the road, there was a checkpoint where our car was searched, and then we drove all night on unlit roads.

We arrived in Culiacán and checked in to a hotel. We had the feeling that all eyes were on us wherever we went – because they were. Not many gringos frequented the city and if they did, they typically didn't move about openly. If we were going to get anything done, we had to blend in. Ski was six feet tall, with pale white skin and red hair, so he wasn't going to blend in no matter what he did. I am Greek, and at the time, was deeply tanned and sported a mustache, so, except for blue eyes, blending in was a lot easier for me. American brand-name clothes were a sure give-away, so I went to a store and bought pants, sneakers, and a cheap pocket knife, and then to the liquor store for a bottle of cheap whiskey. I spilled out most of the whiskey and washed out my mouth with it. People are less suspicious of a wandering drunk than a stranger strolling through their neighborhood. We drove to the area and I got out on foot a few blocks away to reconnoiter the address. I didn't want to walk around with a map, so I had made mental notes on how to get there. The streets twisted and turned with nothing tall enough to act as a point of reference. It was difficult to keep a sense of direction.

Back then, I could have easily run ten miles, but at that moment, blisters from the damn sneakers that didn't fit quite right reminded me with every step of where I was, and what I was doing. In just a few short blocks, I found the address. The street was narrow and busy with lots of people going about their business. As I walked by his house, the man I was looking for was just coming out, and sat in a chair at his door. He looked right at me (I was only four feet from him), so I looked down and just walked/stumbled away. It didn't appear that he was suspicious of me, but that didn't mean he hadn't noticed. I didn't want to look back in case he was watching me, so I walked until I was out of sight. *Damn*, I wished I had gotten a better look at him. I would have liked to be more positive, but I didn't dare make a second pass.

Now to find my way back to Ski. *F*ck*. I was being tailed: one behind me, and the other across the street. I couldn't know if my cover was blown, or if they just intended to jump me. The one across the street sped up and went past me so I was now between them on the street, but before I was caught with no place to exit, I took a left into an alley. It was, of course, a dead end, so I stopped a few feet around the corner, and drew the blade.

It took longer than I expected for the first one to come around the corner, but I can only guess the delay was because he waited for the second guy to get to the alleyway with him so they could come in together. I was flat against the wall and as soon as the first one came around the corner, I stepped out and faced him. He came at me first so I sorted him out hard and fast, and as the second guy attempted his attack, I sorted him out too and then bolted.

I ran a short distance when I realized I was still holding the knife so I wiped it, threw it over a fence, and kept running. Those shitty sneakers hurt, my blisters ripped open, and my heels were stinging like bees. I cursed myself for such a shitty display of fighting technique. I didn't need any training to use a knife, but as a trained empty hands fighter, I knew I had gotten by on violence of action alone.

Life Lesson: Armed is good, but trained with arms is better.

Just then, I noticed my right hand had blood on it and some had transferred to my left hand too. I had an *AHA!* moment when I realized how the term "caught red-handed" originated. I cursed my stupidity again. As I ran, I looked around for a fountain for some water – anywhere to wash it off – but I didn't see any. There was a bottle on the ground with a little bit of some kind of juice so I used that to wash off my hands. It wasn't enough, but it helped. I stuck my hands in my pockets to try and wipe off the rest of it. Ski saw me coming up behind him in the rear view and opened the passenger door. I got in and he asked, "What took so long? Did you find him? What the f*ck happened?"

I said, "Let's get the f*ck out of here, *now*!" and he drove back to the hotel.

Ski asked me what happened and I told him I wasn't sure if I had been made. As I stepped into the shower and the water hit my blisters, it felt like knives were shaving my heels off. While I showered, Ski disposed of my clothes and sneakers. When he got back, I was so thankful that he had also taken the initiative to pick up some gauze and tape at the pharmacy because I sure didn't feel like walking anywhere. Ski went out and got some food that we ate in the room as we watched Mexican TV. It was a long, shitty day; we had no idea if we had blown our cover, and so we slept in shifts in case they came for us during the night.

The next day, we floated the idea of just leaving the city without our photo, but we agreed we'd been through too much to quit, and more to the point, we had given our word to do the job, so we dismissed that idea. We knew we'd get only one shot at getting our photo, so we made a plan.

We filled the gas tank and headed off to finish our mission. We went back to the second-hand store for more clothes, and then Ski dropped me off a little further up the street from our target's house. He would drive around and pick me up a few blocks away. With my head down, wearing a cap, and staying on the other side of the street, I passed his house but he wasn't outside. I went around the block and found Ski and we drove around for fifteen minutes before he dropped me off to make another pass, but this time, I walked back in the opposite direction. Still nothing. *Damn.* I changed my shirt and cap and tried again a while later. Nothing.

One more pass a little while after that and finally, he was in his chair. We drove around and headed up the street so he'd be on the right hand side of the car. I readied the camera, put a T-shirt over the top of it so that only the lens showed, and started filming. As we were driving up on the target, he was holding up a damn newspaper, blocking his face! *F*ck!* But as we got closer, Ski called out "Hey!" and the man lowered the paper and looked up to see who was calling out to him. We got a clear shot of his face. He jumped up and started yelling as we drove off, and someone jumped in their car and came after us.

As we tried to make our escape in the narrow streets, a car came out of a side street and our cars collided, doing a fair bit of damage to the right front quarter of our rental. The driver got out yelling, so of course, we took off instead of stopping, and when his car was blocked, we got away. Ski made a clear path out of the city and we didn't look back.

Now we had 600 miles to go, and knowing we couldn't outrun telephones, we were fairly certain we'd be ambushed somewhere along the way. We even tried to trade the rental for a junker to slip past anyone waiting, but we couldn't communicate well enough with the locals, and the two guys with whom we tried that didn't go for it. We didn't have detailed maps and there was no GPS then, so we had little choice but to stay on the main road all the way back. Much to our disbelief and relief, we made it all the way back to Nogales without incident and crossed the border. Driving back, I was in a philosophical mood, asking myself how I wanted to live out the rest of my life. I decided...

> *Life Lesson: Don't get to the end of your life to find out you lived its length without experiencing its width.*

Vengeance

> "Revenge is an act of passion; vengeance of justice. Injuries are revenged; crimes are avenged." — *Samuel Johnson*

One day, I got a call from the chief of security for the chain of retail stores with which our company had a contract. He told me that a team of thieves had been hitting up the markets for hundreds of cartons of cigarettes every week and he wanted me to put an end to that. Given this top priority, I put my best detectives in different stores with solid teams to support them, and we rotated through the area where stores had seen the most losses.

Finally, one day, I spotted a man enter the store, go to the end of a checkout counter, take a brown bag and put it into a shopping cart, and proceed to shop. Within a few minutes, he went down the aisle with paper towels and toilet paper. He rearranged the paper towels, opened the brown bag and dropped it in behind the paper towels he arranged, and then walked off shopping some more. Within a few minutes, another man who had a shopping cart half full came down the aisle, took two cartons of cigarettes from his cart and dropped them over the paper towels into the brown paper bag and went on shopping. A few minutes later came another guy – same thing. And then a third man – same thing. I alerted the other detectives and told them not to make any moves on them because they were experienced thieves and I didn't want to blow it.

After the thieves had filled the bag with twelve cartons of cigarettes, one guy just took the bag, rolled up the top, and headed for the door. The law states you can't stop shoplifters until they leave the store; just before he walked out, he stopped and dropped the bag right next to the door and turned around. There was another guy waiting outside who was going to grab the bag at the door and take off with it, but he saw me approaching, and didn't take the bag. The guy who had dropped the bag at the door looked at me and started laughing. He said, "You assholes will never catch us," and he walked away laughing. All the men left the store, splitting up in different directions. I watched which way Big Mouth went.

I was furious! Unless we had someone outside ready to pinch their outside man, we were screwed. Thus exposed, and not knowing their tactics ahead of time, they made me so easily, I might just as well have worn a sign. They were professionals who knew the law and devised very clever tactics to mitigate their risk. I went to the store manager and told him, "I am giving official notice that I quit. I just punched out. Please initial my time card." I left the store.

I had watched in which direction the head robber went, so I jumped in my little Fiat, drove around the block, parked my car, and waited for him to get nearer to the corner. When he did, I stepped out in front of him. He laughed at me and asked what I thought I was going to do because I couldn't arrest him. I told him he was right about that, and I proceeded to give him a serious beat-down. Then I told him to tell his friends not to go back to that chain of stores or I'd hunt them down and break their arms. I asked him if he understood clearly and he said yes, so I broke his arm to make sure he got the message. Some guys who were driving past had stopped to watch us fighting and were calling out the action like sports commentators and laughing – until I broke his arm. Then they said, "Whoa! Dude, that was harsh!" But still laughing, they drove off.

I left the team at the store to finish the day and went back to the office to explain what happened. My boss laughed, thanked me for protecting the company by quitting, and "rehired" me. Apparently, during my brief two hours of unemployment, the store manager had called the chief of security and told him the cigarette thieves had been in the store and following a failed attempt to arrest them, I had quit. Deanna, my boss, told me to call the chief and I told him I had screwed up the bust, but that I'd had a conversation with one of the thieves and suggested that he and his friends should shop elsewhere. The chief knew how I operated from other cases, so he didn't need or want any further details. He said he was glad to have someone with my particular set of communication skills protecting their stores and employees. The next week, my boss told me that particular retail chain had added more man hours to our contract.

Life lesson: Sometimes you have to get your hands dirty to get the job done.

Agent of Karma

"Karma uses me to do its dirty work." — *Tom Sotis*

I was having coffee with a city detective who told me that he was frustrated because a custodian at a local old folks' home was stealing from and intimidating elderly ladies, but in such a way that they couldn't arrest him and the department didn't have the resources to set up a sting. He never hinted in any way that I should do anything, and I can't know what was in another man's mind, but he didn't invite me for coffee and relay that story because it was a riveting tale. I got the strong impression that if they had to investigate an assault on this despicable excuse of a human being, they might not dig too hard.

Either way, there was absolutely no way I could let this bully continue to get away with it, so I surveilled him the next morning to watch his routine. On the day after that, I waited for him. It was unlikely I would get caught, but if I somehow did, I would plead guilty and stand firm with pride while receiving my sentence.

"Walrus" (my nickname for him) was quite a bit bigger than me, so I didn't want to just walk up on him in the open parking lot – I wanted to gain an advantage, as attackers always do. I decided to get him when and where he felt safest so that his guard would be down. I waited inside and watched him go down the hallway to his custodian's room. I waited for him to get inside his room and close the door, and then waited several more seconds before I knocked; just as he started to open the door, I body-slammed it with all I had, knocking him back *hard* into the little room. I closed the door behind me.

Walrus recovered surprisingly quickly from the shock, and he was boiling mad. His fury didn't help him one iota because I beat that bully until he cried and begged me to stop. Then I pulled his hair, got right in his face and told him that I *hated* bullies, then *slapped* him in the face *hard* and told him he would return *every nickel,* or I would come back. I also told him that he was now the *protector* of every person in the old folks' home and if I got any bad reports, I would come back. I asked him if he understood, and when he whimpered that he did, I kicked Walrus hard in the balls and left him whimpering, curled up on the floor in fetal position.

I had hurt people before when it was necessary and I didn't particularly enjoy it, but this time was different. I freely admit I felt a strong sense of gratification and to this day, I don't regret it.

Life lesson: Getting your hands dirty is not the same thing as being dirty.

A Lone Wolf

"One enemy can do you more harm than 100 friends will do you good." — *Anonymous*

I had known Andy since I was four years old. He was Greek, our families were close, and he was always Uncle Andy to me. One day Andy called and asked me to stop by his restaurant. He told me that he had put one of his malls up for sale; some guys put down a deposit so he took it off the market while they got their financing squared away. After months of delays and missing deadlines they approached Andy, told him they couldn't get the financing, and that they wanted him to hold the note. Of course, not only did Andy want to sell the mall entirely, but he was rightfully concerned that they wouldn't pay and that he was being muscled out of his property.

Andy gave me the file and I did some digging. The investment group was nothing more than a PO box. I asked a friend in law enforcement to see if my target (one of the men) had a record and it turned out that this man was well-connected in a way you really don't ever want to mess with. At least then I knew who I was dealing with, what to expect, what it would take to get him to back off, and what it would take to resolve it if he didn't. So I made a plan and set things in motion.

Andy invited him to meet at the restaurant and I sat at a different table where I could hear and see everything. The guy showed up and sat down with Andy. He was enormous, like a walking phone booth. Andy told "Hippo" the deal was off, that he would not hold the note, and that he intended to put the mall back on the market right away. As expected, Hippo tried to pressure Andy, saying they would fight him legally, but more concerning for Andy would be the guys in Hippo's home city who would be upset over the negotiations, and that it was better... safer... for Andy to work it out.

I chuckled at Hippo's surprise when Andy told him that since the mall had been off the market for months, the amount of the deposit didn't even cover his losses, so Hippo was not getting any money back: No way, no how, done deal. As expected, Hippo almost popped a vein in his forehead and his face became flushed as the pressure built up from his anger. He was not used to people telling him *no*. But, also as I expected, he didn't do anything right then and there – he was too smart for that. He said to Andy that he was very disappointed and that he'd give Andy some time to think about it in case he changed his mind. And if Andy didn't, he'd regret it.

I got help from two other guys I had worked with before and using all three cars, we easily tailed him home. I positioned them about 75 yards from either side of his driveway and we waited until morning. We planned to photograph Hippo's wife (we assumed/hoped he had one) when she left for work, but I got even luckier. A school bus picked up a child from a house behind the other car and I could see Hippo's front door opening. Perfect! Just as the school bus pulled up and stopped right at the end of their walkway, I pulled up from the other direction and stopped. I had a perfect, unobstructed view and snapped a few photos of his daughters as they waved goodbye to Mommy and got on the bus.

Life Lesson: Having children is akin to putting your privates outside on your mailbox.

I zipped down to the one-hour photo place and had the photos in hand in under two hours. Those photos were about to become a vice-like grip squeezing his testicles. I called Hippo and told him I had a check for his deposit, but that he'd have to meet me that night to get it because Andy didn't want to deal with him any further.

We were to meet at the cafe at 7:00 p.m. My buddies and I got there very early on and staked it out. In case things went bad, I positioned Ronnie inside with me (Ronnie was huge, like Hippo) and two more guys waited outside.

Hippo showed up at 7:00 p.m. and sat across from me at the table. He asked for his check. I apologized for the misunderstanding because I wasn't there to give him his check, that I had something far more valuable to give him, and gave him the photos of his daughters. It took a second for it to register and then I got "the look", the unmistakable look of someone who truly *intends* to kill you. He didn't have to say it.

I knew this was the deciding moment, so I sat motionless, silent, and just stared back. He had to be wondering who had the balls to sit across from him like that and tried to get a read on me, saying I obviously didn't know who he was. I assured him that I knew exactly to whom I was speaking.

> "Just keep in mind: the more we value things outside our control, the less control we have."— *Epictetus*

After another good, long stare, he asked, "Who the f*ck are you? You got some kind of ID?" and I knew I had him. A lifelong criminal, he realized that if Andy had gone to the FBI, I could be an undercover stinging him on an extortion charge. So far, I had done nothing illegal; all I did was inform him that he wasn't getting a check, and gave him photos of his daughters. I never threatened them or him.

Then he asked me, "Are you a Fed?"

Him not knowing who I was gave me all the leverage I needed. I told Hippo I was there to give him a new job ... That someone had threatened *my* family and that now he and his associates were my family's protectors; that if anything, *anything*, happened to anyone in my family, or damage was done to anything they owned, I promised to unleash a Pandora's box of lifelong sorrow upon *him*. I never threatened his family, I didn't have to, and I couldn't be responsible if he inferred something I didn't imply, but nonetheless, my message was crystal clear. I told Hippo he was a somebody with a new job and I was a nobody – a ghost who would disappear and live off his deposit money, just waiting for a call from my family. Hippo never came around again.

Life Lesson: One capable, motivated person can destroy everything you care about.

Extraordinary

> "If you are not willing to risk the unusual, you will have to settle for the ordinary." —*Jim Rohn*

One of my close friends, John Ebert, had become adept at teaching firewalking and most of the members of my training group had done it while I was away on one of my trips to Russia. I am deeply skeptical of any activities that claim to be outside the objective reality that we share, and typically, without verifiable evidence, I suspect them to be pseudosciences.

So when John told me about it, I was dismissive ... not because I didn't believe it could be done, but because I believed there was a trick, as evidenced by the *verifiable* demonstrations of firewalking debunkers who had exposed the scam.

I said, "No way, John. There's a trick. So you're telling me we're going to sing kumbaya and then magically, our flesh won't burn? We are good friends, but that is nonsense." As I always do when presented with such claims I said, "Prove it."

John smiled his big wide grin, shook my hand and said, "Brother, I was expecting you'd say that." And he set up another firewalk.

The evening began with a lecture on the history of firewalking, as it has been practiced for centuries in many cultures. I was curious to find out if they all used the same trick, or if there were different ones. Then the discussion turned to science and about the properties of heat, combustion, fire, and our flesh. John set about debunking the debunkers, and explaining why each of their demonstrations and explanations were flawed and didn't meet the scientific criteria they claimed.

One example showed that we don't burn because of the Leidenfrost effect, which claims your feet sweat, forming a protective layer that prevents us from burning. The Leidenfrost effect only occurs on metal at certain temperatures (like dropping water on a hot skillet) and explains why we can still be burned when we have wet hands. It is an easy experiment; just dip your finger in a glass of water and then immediately touch the hot burner on your stove for just an instant. One by one, John counteracted the theory and demonstration of each debunker.

So, I got right to the very crux of the thing with one simple question and I asked John: If the debunkers were wrong, why then don't they burn?

And that's when John smiled his wide grin again and said, "There are things that science cannot yet explain, but we know and can demonstrate two undeniable facts: one, fire burns our flesh, and two, fire doesn't burn our flesh *when we believe purely and completely* that it won't, and it matters in *what* someone believes, or *why* someone holds their belief. Whether it's to do with religion, science, leprechauns, or unicorns, it has the same result."

Life Lesson: Belief itself, is a force of nature, generated by our will governed and, ironically, by our beliefs.

It all sounded great, but I remained unconvinced, thinking it was all some kind of a ruse so that we wouldn't notice or understand the trick that was about to be played on us.

The fire outside had been burning for a few hours and the huge stack of wood had burnt down to a very large pile of red-hot coals. We went outside as the fire-tender raked the coals out into a large circle about 20 feet in diameter. The heat from just standing around the bed of coals was difficult to bear. John asked everyone to remove their shoes, even if they intended only to observe and not participate. I was one of those people, but of course I complied and removed my shoes. I had no intention whatsoever of participating because I hadn't yet learned the trick, so there was no way I would consider it.

Everyone held hands in the circle and John led us in singing a simple repetitive melody; everybody focused on the song and the fire. Mesmerizing as it was, I remained mentally detached, waiting for the trick. It had to be now. I had my eyes peeled on John and then he let go of holding hands from the two people by his sides and casually strolled across the bed of coals. Wow! Cool trick. Then others who had walked at previous fires started walking too. But they had been before, so they knew the trick. Then, to my surprise, people who had never been before – skeptics like me – walked across the red-hot coals too. And I thought, *what one man can do another man can do.* As I looked up, John was standing across the coals, looking at me and smiling. It is hard to describe, but it seemed like an invisible door was in front of me and I felt compelled to walk through it, and then, with a sense of abandonment, my fears disappeared and in the purest moment of my life, I stepped onto the coals.

> *Life Lesson: Transcending the limits of your physical body requires you to abandon it.*

I crossed the bed of coals in a totally consumed state of amazed empowerment and just as I was nearing the other side, I recall the strange feeling that I was walking on snow. I stepped off the coals, turned around and looked back at the bed when I felt John's arm around my shoulders. I walked ten more times that night.

I began studying firewalking, and after assisting in a solid number of firewalks, I began leading walks on my own. On one walk, I spent 58 seconds on red-hot coals while practicing with live blades (debunking the low-thermoconductivity theory). That night, I traversed the bed several times, and then someone who walked ahead of me accidentally kicked a red-hot coal out of the bed into the surrounding ring and it gathered a thin coating of sand so it wasn't visible. I came off of the bed of coals and stood in the sand, and a moment later, when I was "not of mindful intent", I stepped on that coal (about the size of an almond) and it *instantly* melted the skin off my little toe and a piece of my foot. I was in *amazed agony*. "Amazed" because I had just come off an entire bed of coals with not even so much as a blister, and was all of sudden in agony from stepping on a small ember. Indeed, firewalking is not a "trick" as evidenced by the unfortunate folks who have been severely burned or crippled.

What I learned from firewalking is that our present understanding of physics can't yet explain how *our belief system can alter what we thought to be absolute* (i.e., that fire burns), and proves that indeed, it is not so (i.e., that under certain conditions, it doesn't burn us). Extrapolating that lesson into life, how many other things do we take to be absolute that also may not be so?

Life Lesson: Humans can do extraordinary things when their heart and mind are in absolute alignment.

The Iron Curtain

"You can measure opportunity with the same yardstick that measures the risk involved. They go together." — *Earl Nightingale*

My close friend, John Ebert, not only introduced me to firewalking, but also introduced me to Dan Brulé, the world's leading coach for breathing. In September 1994, Dan had just returned from teaching breathing in Russia. He had fascinating tales of his trip behind the Iron Curtain. Far beyond Moscow, he ventured to the city of Ufa, located in the Central Republic of Bashkortostan deep within the borders of former Soviet Russia. Preceded only by two American Jehovah's Witnesses, Dan was the third American to venture that far into Russia since the wall had come down. I was incredibly curious, and after asking a lot of questions, Dan suggested that I go, too. I asked how I would do that, since I didn't have a single contact, and Dan said to give him my resume; that he would have it translated in Russian, and send it to a girl he knew in Ufa who would circulate it (whatever that meant), and then to get a plane ticket and go. A week later, I gave Dan my resume and purchased a ticket a week after that.

Arriving in Ufa, I was only the fourth American to travel that deep inside Russia. I deplaned and collected my bag on the tarmac. Everyone walked off the tarmac and I found myself standing alone, amazed at where I was. Only if you grew up during the Cold War can you imagine and have an appreciation for what it felt like to be a lone American standing in the middle of "the evil empire" behind the Iron Curtain. I was betting on Mom being right about all those good apples....

I looked ahead and there were no taxis or hotels, just the city. I remembered my grandfather coming to America three generations ago and wondered if this was anything like how he felt once he stepped off Ellis Island. I didn't have any contacts or a plan of any kind, so I decided I would just walk into the city to see what I would find, knowing somehow I would find my way. I was glad I had brought hand and foot warmers and a survival blanket in case I had to be homeless for a bit, but it was cold in November and there was already snow on the ground, so I was hoping I wouldn't have to sleep outside.

As I came off the tarmac, I was approached by a young man who asked, "Are you Tom Sotis?" I was shocked. Affirming that I was, he introduced himself. "Hello, I am Abbas Gallyamov and I will be your translator."

I asked him how he knew I was coming and who sent him. He said the Minister of Internal Affairs contacted the President of the University and requested they send someone who spoke English to greet me. They had selected Abbas because he spoke English fluently and was the region's junior karate champion, so he seemed the logical choice. I guess Dan's friend had circulated my resume after all!

Abbas motioned to a line of six cars and told me to choose one. When I asked why, he said that six of the most influential men in the region had sent their cars. He said it didn't really matter which one I chose because by the end of the day, I would have met all six men. When I asked him why we were to meet them, he asked me where I intended to stay and how did I intend to move about and visit? The purpose was to select a host who would care for me while I was there. I couldn't believe it: I was nobody, and they were treating me as though I was somebody. It felt very strange and I wondered if it was some kind of a set-up or trap, and of course, I had little alternative except to go along and see what happened.

> *Life Lesson: Adventure happens when curiosity outweighs fear of the unknown.*

After meeting with five of the six very gracious men who offered to be my host, we had the final meeting. I was impressed with the way Vladimir L. carried himself, the way his men carried themselves, and the manner in which they received me. Always calm with a pleasant demeanor, Vladimir was a tiger wearing human skin and was the obvious choice for me. When Vladimir offered to be my host, I told him I was honored to accept. He asked if I was willing to train his men for two hours every morning, to which I gratefully agreed. After a nice dinner, Vladimir and Abbas took me to a flat. They said Abbas would come the next day to collect me, gave me the key, and left.

I looked out the window into the city. Then the full realization that I was alone, 10,000 miles from home, behind the Iron Curtain enveloped me and I was *exhilarated*! The unknowns ahead made me feel like I was balancing precariously on the edge of a mile-high precipice.

First Impressions

> "You never get a second chance to make a first impression."
> — *Anonymous*

In 1994, Russia was still finding its feet amidst the chaos and turmoil of social and economic restructuring. At the time, the government was battling with the mafia for control and protection, and security companies were aligning with either side. It was an ultra-violent era in recent Russian history and there were open gun battles nearly every day. Employing hundreds of men, Vladimir had the largest security company in Ufa and was aligned with the government. The first day of training went well. They seemed pleased with what I shared with them and I mistakenly imagined the rest of the trip would be pretty much the same.

Then, Dan Brulé arrived, and being so far from home, it felt great to be with another American. We met up after the morning workout and spent the day together, meeting friends Dan had made on his trip. After dinner. Dan headed off to wherever he was staying and we planned to meet up again the next day. I was with one of Dan's friends who asked me if I would be so kind as to grant him an interview. I was happy to oblige.

We arrived at a small building, with a room that had a couch and a TV camera. Doing the interview was fun. When we finished, the host asked me if it was OK to play music, which of course was OK with me. So he pulled out his guitar, handed me a mouth harp, and began to play himself, motioning me to play the harp. I was gesturing, as if to ask: "What do you expect me to do with this?" and he just kept playing and nodding for me to join in. So I just laughed, put the harp in my mouth, and strummed away, completely out of tune. Finally, he stopped playing and I stopped too, glad the humiliation was at an end. The host thanked me and asked if I would be so kind as to do him and the audience a small favor before concluding. Of course, I agreed – I wouldn't refuse a small favor to a gracious host on TV. Besides, I figured we were done so it must be something simple and easy. I figured wrong. The host grabbed his guitar, handed me a sheet of paper, and said, "Great! I even wrote the words down for you," and he began to play.

Trust me when I tell you that if there is a worse singer in the world than me, they are keeping it a secret. I looked at the words and to my shock and dismay they read: "Love Me Tender", by Elvis Presley. You had to be kidding me! So, I opened my mouth and words came out that I dare not even call singing. At least it was a tiny station and hardly anyone would see it. Wrong again. It turned out to be Bashkortostan's only TV station for over a million people and they ran the interview a couple of times/day for a week! My humiliation increased by a factor of ten.

After that, I was very highly recognized pretty much everywhere we went. Just walking down any street people would call out, "Hey, American!" from their vehicles or from buses. It turned out that the Russian people took somewhat of a liking to me because I was willing to be embarrassed and play along, that I didn't think myself superior to them. They say you never get a second chance to make a first impression, about which I learned something unexpected.

Life Lesson: Sometimes the best way to make a first impression is not in a way that you would prefer.

Bravery Wears Disguises

> "What you have outside you counts less than what you have inside you." — *B.C. Forbes*

On the third day of training in Russia, I arrived at the gym. Abbas (my translator) was talking with Vladimir and I noticed Abbas' face turn pale. I asked what was wrong and he said the Minister of Internal Affairs was coming to meet me for a personal demonstration, and of course Abbas was to translate.

The minister arrived, and after a formal introduction I gave him a 10-minute demonstration, which they filmed. When it was over, the minister said that my training with Vladimir had been scouted by the Spetsnaz (Special Forces) who reported that I did things "differently" and the minister invited me to go to the Spetsnaz training facility for further evaluation. When it comes to fighting, "invited for evaluation", means the same thing in every language. I responded by asking if it would be too much trouble for them to send a car to collect me because I would never want to keep anyone waiting should I have any difficulty finding the location, and they were happy to oblige.

On the way back to the flat that day, I was in a good mood. Abbas was not sharing my outlook and when I asked what was wrong he said, "What do *you* think is going to happen tomorrow?"

I said, "I expect to get my ass kicked."

And he asked, "So why aren't you more concerned?"

And I said, "Whatever happens, I'll earn their respect." Again, most things in fighting transcend language....

Of course, no mentally healthy person looks forward to getting his or her ass kicked, but there was never really a decision regarding whether to go or not. If I didn't accept their invitation, I might as well get on the next plane home to live every day of my life thereafter with the shame of knowing that I'd let my country down. I realized that perhaps nobody may ever know what was about to transpire – that I could very easily just disappear, remaining an "unsolved mystery" – but *I'd* know, and that's all that mattered. Like most other average Americans, I'd rather have died right there on the spot than be a coward when my countrymen would expect me to be otherwise. As I lay in bed that night, I could hear my mother reminding me they are sons, and brothers, and fathers, too ... just like us.

The next day, the regional commander of the Spetsnaz came to pick me and Abbas up at the flat, and when we came out, Dan was waiting to go with us. I pulled Dan aside and told him it was a dangerous day; there was a good chance I might not come back, and it was probably best if he did not come along. Dan scoffed at that idea and said if another American was in danger, he'd never desert them in a hundred years, and he went and sat in the chief's car. It turned out that the easy-going breathing guy had courage.

Life lesson: It is a mistake to estimate someone's courage by their everyday demeanor.

The Spetsnaz

"Once I dared to dare greatly." —*Morgan Matson, The Unexpected Everything*

The Chief of Spetsnaz put the police siren on, and drove us straight to the Spetsnaz compound on the outskirts of the city. Two massive metal doors opened and we drove in. Everything was dirt, wood, cement, or metal. There were three noticeable colors: grey, greyer, and old grey. Abbas mentioned I was the first American to be on a Russian military base in Bashkortostan, ever.

I felt good as I stepped over the threshold of the metal door into the gym, knowing that whatever happened, it was OK because right there, I was making my mark. I was 5,000 miles from home and challenged by some of the toughest guys behind the Iron Curtain. *I'd showed up.* At least I had done *something* with my life.

We entered into a gymnasium with about a hundred guys, shirts off, more abs than a rolling-pin factory, and they were drilling kicks into each other's stomach. Every guy in the room was a rock-solid warrior who had proved their mettle in one of the most physically and mentally demanding selections in the world. I didn't know what their intentions or instructions were, and for the first time, I really felt my life was in jeopardy. Maybe their Russian mothers hadn't given them the same life lesson about Americans ... I expected that before I got out of there, *if* I got out of there, that at least one of them might try to kill me, and even if that didn't happen, it was surely going to be a long and painful day.

When we entered the second gym, there were about thirty people sitting in a group in the bleachers. They introduced me to the group and introduced the group to me. I was in the company of some of the highest ranking military and government figures in Bashkortostan.

I remember when I first saw the group of men they had assembled for me to fight. I admit they were an intimidating group and I could see in their faces there was no lack of pressure on them either. It felt as though there was an entire migration of monarch butterflies in my stomach, and my adrenaline was already leaking. I could hear my grandmother's words: "Just two hands and two feet" and I remember thinking, "But Grandma, some of them have really *big* hands!"

We fought with weapons and empty hands. I fought with two sticks, one stick, stick-and-knife combination, and with knife against a variety of their preferred weapons, which were mainly traditional martial arts implements. I matched empty hands with their martial artists and boxers. If it is fair to say I had an edge with weaponry, it is equally fair to say they had an edge with empty hands. Their Sambo fighter was a machine of skill, speed, and power who made me tap out pretty fast. I guessed if someone had been instructed to break my back that day, it would have been him, so that was a sign I might get through it after all.

If I had to pick one day in the last fifty years of my fighting experience, that day holds the most significance for me. I will live all my days knowing I presented a worthy example of the

American fighting spirit. I am glad Dan Brulé and Abbas Gallyamov were present to witness and confirm the events of that day.

At the conclusion, the regional newspaper interviewed me. Apparently, they liked my responses because they called me "the American with a Russian soul" and compared me to a famous Russian poet. Those newspaper articles are among my most cherished possessions.

Afterwards, Abbas, Dan, and I were treated to traditional Bashkiri hospitality (vodka shots and snacks) and when the Chief of Spetsnaz joined us after their meeting, he asked if I would train some of the teams, and with that, I became the first American private contractor to work in Russia.

Back at my flat, I was glad my two good friends were waiting: Advil and Tylenol. I remember looking out the window at the city lights and thinking, "You were right, Mom, again. There's good people everywhere."

> *Life Lesson: When you have pure intentions, you do not fear the outcome of your actions.*

The Mark

> "The robbed that smiles steals something from the thief." — *William Shakespeare*

One time in Russia I ran out of rubles, so Abbas and I went to the local hotel to exchange $100 USD. When we arrived, two guys approached us in the lobby and asked if we wanted to exchange money. As we were about to make the transaction, another guy walked by and warned us about them; one of the two slapped him in the head and chased him off.

The money-changer wanted to inspect the $100 bill to be sure it wasn't counterfeit. As he was doing that, the guy who had been slapped came running back at them, hit the guy who slapped him, and ran out the door. Quickly, the money-changer handed me back my $100 and the two of them gave chase. We decided to leave right away and as we walked out of the hotel, I looked in my hand and found a $1 bill. The whole commotion was a set-up and I had been ripped off. I started laughing and Abbas asked why. When I showed him the $1 bill he was surprised too, and angry. He asked why I wasn't angry. I told him that, "learn-by-experience survival lessons are usually costlier".

> *Life Lesson: Sometimes you lose a little now so you don't lose a lot later.*

Just Because

> "We hate some persons because we do not know them; and will not know them because we hate them."— *Charles Caleb Colton*

Just outside the hotel was an area with tables where people drank tea. As we got closer, Abbas turned me around and said we should go the other way, and when I asked why he said, "Those men with thick beards: they are Georgians." And when I asked why that was an issue Abbas said, "They will kill you." When I asked why again, he said, "Just because you are here."

> *Life Lesson: Some people may want to kill you because you represent or symbolize something they hate.*

Little did I know that would become a repeating theme in my life.

Visionaries

> "Nobody but you have to believe in your dreams to make them a reality." — *Germany Kent*

For many years, I had dreamed of having my own martial arts school, but once I was living that dream, I quickly became discontented when it became routine. I also realized that no matter how much I might advance in the martial art that I was doing, "martially-speaking", I would only ever be just another brick in the wall. If I was going to do something interesting and challenging, I would have to specialize.

I researched every martial arts magazine I had in my extensive collection and bought every new issue that came out. In them were listings of seminars, so I began to track them to see who was popular on the circuit and what were they teaching. At the time, Jeet Kune Do and Filipino martial arts were gaining in popularity and both had a "streetwise" element to them, so I created Analytical Combat which combined my unique 4 Themes Theory with elements gleaned from street-fighting in Los Angeles.

I began teaching Analytical Combat seminars at local schools where I already knew the instructors. But when it came to introducing the new approach to instructors I didn't know, I soon discovered that most martial arts instructors suffered from such fixed ideas, and resistance from cognitive dissonance that it was harder to teach them than it was to teach people with no previous training. At least the minds of the uninitiated were open to learning. What I also learned was that I loved teaching via seminars.

Seeking to specialize, I noticed very few guys who seemed to be openly teaching knife fighting. These instructors kept it close to the vest and very rarely taught open seminars. Some Filipino martial arts masters taught knifework, but only to their most dedicated students.

Filipino martial arts, typically thought of as stick-fighting arts, also had knife work in them, but it was "behind the rattan curtain", taught only to family members, rarely to outsiders, almost never to non-Filipinos – and in those cases, usually not the complete system.

> "The question isn't who is going to let me; it's who is going to stop me." – *Ayn Rand*

When I researched why nobody was teaching knifework openly, most had a very limited amount of knifework. Also, they didn't want their own knifework to be known by others if challenged by them, and some thought it to be socially irresponsible.

My thinking on these points was that I was willing to accept challenges; I would study widely so I had plenty to share, I was unafraid of others knowing what I teach – even challengers – and I would teach in a socially-responsible way.

I spoke with each of my mentors about my vision to teach knifework openly in seminars, and, without exception, not one thought it was a good idea. For one reason or another each of them thought it would lead to a bad end. I was, and remain, grateful for their advice.

Nonetheless ... I had a vision, and even if nobody else could see it, I wasn't going to let their lack of imagination prevent me from making it a reality. My vision was/is to normalize the use of knives for self-protection from criminals.

Life Lesson: Don't permit others to obscure the clarity of your vision.

The Rattan Curtain

"Trouble is the place where you find yourself when your judgment malfunctions." — *Bangambiki Habyarimana, The Great Pearl of Wisdom*

I aimed to study bladework with Leo Gaje Jr., but he had gone underground for eight years. So my teacher, P. Greg Alland, who had been one of Leo's top students, arranged for us to go to the Philippines to study with Leo's uncle, Jerson "Nene" Tortal, who was also headmaster of their family's martial art. We timed our training camp with Nene to coincide with the World Kali (stick fighting) Tournament and with hopes of meeting Leo, as it was rumored he sometimes attended.

On our flight from Boston to the Philippines, we had a layover in Los Angeles and our group was joined by another practitioner of Filipino martial arts, David Gould, who would soon become a pivotal figure in my life. After the long flight, we arrived to a warm greeting by "Tuhon Nene," who took us to meet the region's superior court judge (or his equivalent) who graciously invited us for tea to welcome us to the Philippines. He told us he, too, was a Kali fighter and recounted some stories of using Kali to protect himself. He asked if we had knives and if we knew how to use them. We said yes and he gave his business card to each of us. The judge told us, "If you encounter danger, use your knife to protect yourself. Then give this card to the officer who responds. He will take you directly back to me, and I will return you to camp." That was an unexpected and interesting welcome that made an impression on us. We proceeded to the Mambukal Mountain camp, had dinner, and talked for a couple of hours before retiring.

The next day, the tournament began with the Philippines National Anthem followed by various speakers, and before the fighting began, various masters gave demonstrations to the crowd showcasing their arts. Most Filipino martial arts schools specialize in one or two weapons, but nearly all train with a variety of weapons, including, but not limited to: one stick, two sticks, stick and knife, knife, two knives, swords, as well as empty hands. I was in combat heaven marveling at the skills exhibited by some of the world's best weapons fighters.

Then, the Leo arrived! I wanted to meet him, but I didn't want to be introduced and be just another person in a line of people trying to get a minute of his time. So I paid the announcer to call my name, and with one of my students, went on the floor. The announcer told the crowd that he had checked our blades and that they were razor sharp. We proceeded to surprise them with a dangerous, high-speed, live-blade demonstration that succeeded in capturing Leo's attention.

After our demonstration, Nene introduced us to his nephew, Leo (they are actually only two years apart in age) and after they had a discussion, Leo said he would meet us back at the Mambukal Mountain camp. This was great news because I hoped that he might also train us along with his Uncle Nene. The day was starting off great!

Because we were foreigners with no ties to any of the schools in the tournament, they asked us to officiate the fights to ensure fairness. Just being there was cool, but being asked to judge the fights was an unexpected honor for all of us. Even though they wear protective equipment, stick fighting is brutal, and there was no small amount of blood spilled that day. Fighters from different schools would match up and fight as though they were trying to end the other's life, and when it was over, they would shake hands and often go talk to thank each other for an opportunity to fight their best.

Jet-lagged and exhausted from the heat, it was a long day that lasted past midnight. We were happy there was a food cart that served chicken and garlic rice so we sat on the curb, snacking, until the taxi arrived to take us back to camp. There were eight of us, so we took two taxis back up a long, winding, dark and mountainous jungle road. Quite a way up the mountain we could soon hear faint music that grew louder as we approached, and to our surprise, we came upon a football-sized field with a live band playing and hundreds of people drinking and dancing. And there was a shack where they were selling beer!

Fourteen hours in the stifling heat with insufficient water had obviously impaired our judgment because we decided to see if we could purchase a case of beer. The taxis pulled over and I got out while the group waited in the cars. I made my way through the crowd to the shack and noticed I was getting some unfriendly looks. The guy behind the service counter just stared at me, not saying a word. I was very apologetic for intruding and said I meant no disrespect in any way, I was just wondering if they might be willing to sell me a case of beer. He gave me the universal "f*ck off" look and within those few seconds, I had a semi-circle of guys standing around me. I apologized and was surprised they let me walk away, but they followed me until I got back to the taxis and then they melted back into the crowd.

I told the guys it was dangerous and we should hightail it out of there, but they said we couldn't leave because instead of staying together as instructed, Kenneth – one of the guys who had come with Greg – had wandered off somewhere into the crowd. I told the guys to stay together while I found Kenneth and I waded back into the crowd. It took only about two minutes, but the pack of guys following me seemed to be steadily growing and I started to wonder if I was going to get back out in one piece. Then I saw a small clearing by the stage and it was because Kenneth was standing there, staring at the band, while everyone around him backed away. He was standing there alone and oblivious to what was going on around him. I grabbed Kenneth by the arm and dragged him back to the waiting taxis, expecting every step of the way that the group of guys would descend upon us. But it didn't happen. I pulled Kenneth through the crowd out onto the street and as we crossed the street to the waiting taxis, I saw all the guys were out of the cars. There were also Filipinos on the other side of the cars, and as Ken and I got to the taxis, our group of eight was encircled by about forty guys.

We could see that things weren't going to go well for us, so we decided to make it not worth their while. I quietly told the taxi driver that if he didn't get us away when we jumped in the car, I'd kill him first. On cue, we all drew our blades and I declared our readiness to fight to the death, and that we knew we might ultimately lose, but we would take many of them to the grave with us. I asked, "Who wants to die first?"

We surprised them with our unified aggression and they hesitated in deciding what to do, so we jumped in the taxis and barely made our escape. We sped into the darkness while they chased us up the road, throwing bottles at us. It was indeed a long first day.

Life Lesson reinforced: Some people may want to kill you because you represent or symbolize something they hate.

The Extra Mile

"You know you're at training camp when you wake up and all you can smell is BENGAY®." – *Guro Dave Gould*

On the first day of training in the Philippines, I awoke at 4:00 a.m. because of monkeys barking in the trees outside my room. Strangely enough, during the night, someone came into our room and painted a one-inch black stripe along the wall, onto the table between the beds, up onto a soda bottle, down the table leg, and out the door. I rubbed my eyes and when I looked closer, I noticed the stripe was *moving*! ANTS! I flew out of bed. An army of ants had made their way to the soda in the bottle, and then out the door.

Each day started by hydrating and standing at attention on the field at 5:00 a.m. and you never wanted to be even ten seconds late. Every morning began with two hours of footwork drills: first linear and lateral movement, then angular and circular stepping, followed by zoning and shirting with variations on crouching, squatting, kneeling, and so forth. After two hours of cardio, we had breakfast which sometimes consisted of a couple of eggs with fruit or sometimes cold fish with ants on it. We guessed it depended on which cook prepared it and how much they disliked us.

After thirty minutes for breakfast, the next few hours were hard drills in the blistering sun. The field that we trained on seemed to have zero breeze except when it was treating us to a whiff of a dead, decaying dog, or strong odor of sulfur that somehow emanated from the earth. Everyone vomited nearly every day.

Sometimes we stood motionless in poses, holding heavy swords, and other times we just made thousands of strikes with our sticks. There were days where our blisters got so bad that we'd drain them and then duct tape our sticks into our bloody hands so we could finish. Then we had an hour for lunch, which was barely enough time to peel the stick from our hands, wash up, change our socks and shirts, quickly eat, and be back on the field.

We had a love/hate relationship with the afternoons because that's when we finally spent time doing tactics training, but the heat could have melted a rock. We practiced attacking, countering, re-countering, faking, probing, and recovery drills while *roasting* in the blazing hot sun for four more hours. Five o'clock was always a welcome reprieve from the tortuous afternoons. We'd enjoy a shower, and the evening meal was always better than breakfasts and lunch.

We'd finish dinner at 6:30 p.m. followed by an evening light workout, sometimes slowly reviewing the afternoon lessons;

other days, we had theory lectures until 10:00 p.m. Exhausted, we would fall up the stairs into our beds and be fast asleep before our heads hit the pillows.

> "If you want to take the road less traveled, you better be ready to go the extra mile." – *Tom Sotis*

Dreams Made Real

> "Life imitates Art far more than Art imitates Life" — *Oscar Wilde*

Jerson "Tuhon Nene" Tortal is the epitome of a patient and caring mentor, sincere about developing you. That is not to say he isn't also demanding and hard on your mistakes – it means that he is a genuine and honest man who shares generously from his heart. And it is a sad commentary to say that quality is indeed both rare and refreshing to encounter.

Immediately, I felt a close kinship with Tuhon Nene; he reminded me of my dearly departed grandfather. When Leo decided to join our camp after the World Kali Tournament, Tuhon Nene offered his room to Leo and his wife, and I was honored that he requested to room with me.

As if the grueling, seventeen-hour training days weren't enough, every night for the first week, Tuhon Nene woke me up and asked me to show him my knifework. He never said (nor did I ask) *why* he wanted to see my knife-work ... I assumed that maybe he was picking my brain. But the far more likely reason was that, apart from testing their training material, he wanted to analyze how I thought and moved.

It is a fairly common theme among martial artists to daydream about journeying back to the motherland of one's chosen art to study with an authentic master, but only a few manage to make it happen. Awakened from a sound sleep to do knifework with the master in the dim light cast by the bulb outside the window always felt dreamlike and surreal.

Toe-to-toe in close quarters, we were striking, parrying, hooking, and blocking, while deflecting, checking, posting, and monitoring. Sometimes, Tuhon Nene would let me take the lead in a drill, thus "dictating the dance" and then he'd purposely throw me a curveball to test my reflexes apart from the actions delivered deliberately. Working with Tuhon Nene in the dim light of that cramped room, I was very fortunate and grateful to gain valuable insights both from what he shared with me, and how he did it.

I never asked why we didn't just practice before bed and why he'd wake me up; I just figured he had his reasons and trusted that he'd tell me if it was important for me to know. Bad to the bone, this particular master had a special kindness and I became quite fond of him. Tuhon Nene became like a favorite uncle and I cherished our time together. Strangely enough, even though I was mentally exhausted, those surreal midnight sessions are burned vividly into my memory.

Life lesson: Sometimes reality imitates your dreams.

Surrounded by AK-47's

Me: Why did they want to kill us?

Leo: Because you are here.

On another trip to the Philippines, we were training with Leo a short distance away from Mambukal, in Buenos Aires, Negros. Back to seventeen-hour days again. One night was just too humid to sleep, so I got up at 2:00 a.m. and went downstairs to find Leo and three of the Dogeaters arming themselves with swords and knives. I asked what was going on and Leo told me there was a problem outside and they had to take care of it. I grabbed my sword and knife and went with them, but when we got to the door, Leo said I could not come outside. When I asked why not he said, "Because *you* are the problem and it will be worse if you are present". I protested and he permitted me to go outside, but instructed me to remain at the door and not to enter the courtyard.

The four men went about fifteen paces to the middle of the courtyard and Leo began speaking loudly in Tagalog. Just then, the bushes surrounding the courtyard moved, and instantly, we were surrounded by guys with AK-47's. One of the men approached Leo and they had a heated discussion. At one point, Leo put his hand on his sword, and the guy held up his hand so none of his guys would shoot. They each said another sentence or two and then those men vanished into the blackness of the jungle as quickly as they had appeared.

It turned out that they were Huks, and they'd heard Americans were there and came to kill us. (Huks are the remaining insurgents of a defunct socialist/communist guerrilla movement). Leo told them that we were his guests, and the Huks would have to kill them to get to us. The Huks said they weren't there to hurt other Filipinos, so they would let it go, but warned that once we left, if we ever came back, that next time they would kill everyone, Filipinos included.

Life Lesson reinforced: Some people may want to kill you because you represent or symbolize something they hate.

I remember vividly how a single spotlight lit the courtyard, and diffused by such dense humidity, created a fog-like scene in this tiny jungle clearing of four men with swords, facing off with a dozen men armed with AK's: another surreal experience burned into my memory.

Leo instructed me not to tell the other guys until the end of the trip because he didn't want them distracted during training. The next day though, my guys were asking Leo why he and some of his guys were wearing sidearms. He smiled and said, "Well, after all, we are in a third world country." The Dogeaters just looked at me and smiled. At the end of the trip, the Dogeaters told my guys the story and their mouths fell open as they thought about how they slept through all of it.

One Foot in the Grave

> "Arrogance is an unhealthy ego in need of repair."—
> *Thomas Franada, The Global Sales Professional*

On one trip to the Philippines, a guy named John, who occasionally trained with me in my early days in karate, wanted to attend. I told him I didn't think it was a good idea for him to come along because he had an attitude problem that would not bode well among Filipino warriors. He assured me that he would be on his best behavior. I gave him clear and fair warning that if his big mouth got him into trouble, I would not risk myself or any of the guys to get him out of it.

On one of the ordinary, blistering, oven-hot days, we were training on the open field with rattan staffs about five feet long. One of Leo's assistant instructors, Baldwin, was putting us through drills and he kindly gave us a water break. John had previously studied staff fighting with a renowned Okinawan master and fancied himself skilled.

As we headed off the field into the shade for a drink, John hit Baldwin's staff with his own in a semi-friendly, challenging manner. It was foolish because there are no friendly challenges, just challenges veiled as such. Baldwin was rightly offended and eagerly agreed, but Leo came out and told us to finish training, so Baldwin told John they could cross sticks that evening. After dinner that evening, Baldwin was not among us and I wondered where he was because I knew the incident was not forgotten – on the contrary, it festered.

That night, when everyone had gone to bed, I was spending some time with Leo when Baldwin came in with Rommel. They were heading upstairs when Leo called them over and asked what they were up to. They told him they were going upstairs to get John to bring him out to the field to cross sticks. Leo castigated them, saying it would spoil the camp and be bad for his reputation if John disappeared, so they had to drop their plan. I have good reason to speculate that earlier on, John had convinced Leo he would invest in one of his schemes, and that is perhaps the only reason he survived the night.

Life Lesson: Arrogance coupled with ignorance could cost you your life.

Leo told me not to tell John because he didn't want anything more to come of the incident. The next morning, I pulled John aside and told him to watch his step because he had one foot in the grave and the other on a banana peel.

The Ceremony

"Go confidently in the direction of your dreams. Live the life you have imagined." — *Henry David Thoreau*

I went to the Philippines for two main reasons: to train with an authentic master, and to get a measure of my own knifework. At the time, Greg Alland was my teacher, but I was still an independent practitioner in self-development mode, with no standard yardstick by which to measure my work. When blazing your own trail without a map, it is easy to do a lot of work in the wrong direction, or even go in circles. I wanted to train with the best to determine if I was on the right path, or off on a tangent.

At the end of our training, we were invited to a dinner where there was a formal ceremony complete with swords and candles, and were adopted into Nene and Leo's martial family. Each of the masters and instructors gave demonstrations, as did some of us who attended. I am glad I filmed everything because the demonstrations were exceptional, and I captured what will someday be historical footage for that family's martial legacy. While a number of foreigners may achieve rank in their fighting systems, few are adopted into the family, and only then receive held-back teachings.

The Dogeaters warrior tribe are a group of local Kali men. Most are students of Leo's, plus a few who trained with other masters too. I was the first non-Filipino accepted into the Dogeaters, with whom I remain close friends to this day. Initially, the

Dogeaters were not unfriendly, but they weren't particularly welcoming either, and for good reason. A vast majority of martial artists who do travel to the mother country and study with an authentic master are seeking photo opportunities and rank promotions more than knowledge. But the Dogeaters hard-earned every scrap of knowledge with sweat, blood, and pain so until you prove that you are cut from the same cloth by subjecting yourself to sufficiently rigorous practices, they don't waste their time patronizing ass-kissers. That the Dogeaters eventually accepted me for my dedication, in spite of my nationality, was deeply meaningful, and twenty-five years later, I am always happy to see they are also traveling the world, teaching. We still communicate regularly.

And that wasn't the end of the evening. It turned out that Tuhon Nene had awakened me each night because he was evaluating my knife system, and based on his recommendation, I became the first foreigner to have his knife system recognized by the Bacolod Negros Arnis-Kali Escrima Federation (the organization of the Grandmasters of Filipino martial arts). An American recognized as headmaster of his own knife method... I was stunned. I have never cared about ranks and titles because they are artificial, but this recognition was a validation of my work, and it cemented my mission to widely teach knifework.

Life Lesson: Sometimes dreams do come true just as you imagined.

Guarding the Legacy

"The glue that holds all relationships together — including the relationship between the leader and the led – is trust, and trust is based on integrity."— *Brian Tracy*

On my next trip to the Philippines as a family member, I hoped to receive previously held-back teachings, but you never know what it may still take to earn them, or how long the process may be. If you aren't fully committed to training without any guarantees except more pain, don't even bother. It was widely known that Leo had a complete knife system because even though he had stick-work, he had a greater reputation for his knifework among Filipinos.

On that trip, we met Leo in Cebu. The guys wanted to be able to see more of the country, so Leo agreed we would train only from 6:00 a.m. to 6:00 p.m. The guys would have the evenings free, and he would move us to a new location every three days. It turned out that Leo had not just one system of knifework, but three, each prioritizing and emphasizing different tactics and elements giving rise to their uniqueness. Previously, I had learned the first system in its entirety, so Leo taught me both of the other two systems at that camp. They were relatively easy to pick up because they were based closely on the knifework from the main system.

During the evenings, when the guys were otherwise occupied, I spent a lot of one-on-one personal time with Leo. We talked about many things and he shared a lot of his life experience with me. On the last night of camp, Leo sat me down for a private discussion where he demanded that I stop teaching my own method of AMOK! and just teach his three methods.

I thanked Leo for the opportunity, but I told him I would never do that because then he would never be able to trust me again. I explained that most others seek him out because they wanted something from him and his family's legacy. They wanted information, rank, titles, recognition, positions within the organization, etc. They wanted a piece of him. And someday, within a week after he passed, they would fracture his system into a dozen systems, each making up a bunch of stuff they'd claim he taught to only them.

I reminded Leo that when I had begun training with him five years' prior in Mambukal, I'd made it clear from the very start that I desired no rank or titles or martial affiliation, that I aspired to be only a student and just a student, remaining at the bottom of the ladder. And because I had AMOK!, I had no need for his family's knifework, or anything else for that matter, and if it suited him, I would return every photo we ever took and every piece of paper that had passed between us. I thanked him for the opportunity to show him how much he could trust me.

Life Lesson: Anything you are willing to sacrifice cannot be used as leverage against you.

Not only did Leo have no leverage, but more importantly, he knew I was right about the trust issue. The next morning, Leo requested I get the video camera and together, we filmed all three of his family's knife systems. Leo stated on the film, in case he should die before he could train his own son, that he had appointed me guardian of his family's three knife systems. I am the only person he trusted to protect and preserve them for his son. I gave him my word I would never teach those systems to anyone except his son.

Life Lesson: It is easier to trust someone who doesn't profit from your relationship.

The Clown and the Prostitute

"Blood spills an undeniable truth." — *Tom Sotis*

For many centuries, the very nature of martial arts produced a self-regulating standard. The best fighters/methods survived; the others died along with their methods. Posers and buffoons dared not venture into this realm where the penalty for exposure was certain disability or death. *Legal liability* dismantled this centuries-old means of self-regulation and the entire martial industry became unregulated (and, I believe, should remain so).

Consequently, among many various martial arts throughout the world, a number of governing bodies (associations of grandmasters) formed which attempt to establish and regulate legitimacy. The World Headmasters Council (WHC) is one such association and many martial artists seek their official recognition in validating a method they devise. Such councils are political by nature so, personally, I have no use for them.

Eugene Coutu, my instructor in Kenpo with whom I maintain a strong relationship, was, at the time, on the governing board of the WHC. A man named Richard Petronelli, a member of the council, brought his friend Eddie Villareal before the council where Villareal made a formal request to have his art recognized. Villareal claimed he was a legitimate master of knife fighting under Leo Gaje Jr. in the Philippines, and that he had learned Kuntao from Petronelli, and having merged those two authentic martial arts into one, Villareal wanted to have his art formally recognized as legitimate and authentic. The members of the WHC admitted to knowing little about knife fighting and didn't know if Villareal was the real deal, but they knew my reputation, and they knew I was Gene's student, so they asked Gene to ask me if I would investigate on their behalf.

Coincidentally, I was returning to the Philippines the following week and while I was there, I interviewed Leo on video with regard to the matter. Leo stated he had never met Villareal, that Villareal was a liar who was attempting to steal his family's legacy and that the WHC should summarily dismiss him.

When I returned to the USA, I played the video tape for Gene. When Gene told the council that my investigation revealed Villareal to be a liar, Villareal protested, claiming I had never even met Leo and that I was lying. Gene felt his word and mine should be sufficient, so he did not offer the videotape as evidence. Petronelli suggested the matter be settled by combat, and the loser, if he survived, would retire from teaching for life. Gene knew I would expect him to accept on my behalf, so he did. Villareal and I were to meet at a time and place appointed by the WHC, and duel.

The duel was set for the next Sunday afternoon at Petronelli's martial arts school. I was risking both my life and liberty so I met with an attorney to discuss the aftermath of the event. On Tuesday, Petronelli called to tell me he was concerned about liability at his school and that the duel was canceled until the WHC set another date, time, and place. On Wednesday, Villareal called to tell me he intended to kill me anyway, and that he was also going to kill my family. Then he called Gene's karate school and left that same message on the answering machine, telling him what he intended to do to me.

That next Sunday, I had a folding knife in my pocket, a bag with a hardwood fighting stick, and the dueling blade that belonged to Leo's grandfather. Dan Neilan and Randy Hodges were my top guys, so they picked me up and we headed off to Petronelli's school. They asked me why we were even going since it was canceled. *Because of a lesson I learned long ago: once you give your word to fight, you show up no matter what.* When we arrived and I saw a lot of cars in the parking lot, it fueled my suspicion that Petronelli had set me up; when I entered the school and saw Villareal, it was confirmed.

When I walked in, Villareal's jaw fell open and everyone in the room fell silent. Petronelli was standing in his office, frozen, jaw also hanging open. I asked Villareal to explain why he was there, and why was everyone else there, when Richard had told me the duel was canceled. I said they were all invited to witness what Petronelli and Villareal thought would be my no-show to a challenge so they could say that I chickened out. They staged the entire thing, and it had blown up in their faces.

Desperately, Villareal tried to launch a surprise attack with empty hands strikes which I easily warded off. Then Villareal lifted his shirt, reached in the waistband of his pants, pulled out a length of chain with metal weights on each end, and swung it at my head. I used my left hand to deflect the chain and it passed overhead.

Then I saw "the look" in his face and eyes and I knew he was going to come at me with everything he'd got. He took another swing with the chain at my head, but this time, I extended my left arm up a bit higher, so instead of passing overhead, it wrapped around my arm. Simultaneously, with my right hand, I drew and deployed my tactical folder. As I pulled him in with the chain, I stopped the blade one inch from his heart.

He stood, frozen in shock for a moment. Then he dropped the chain, ran into the karate room, grabbed a staff off the wall and came back. Without taking my eye off him I extended my hand behind me and said, "stick" and Randy put my fighting stick in my hand. I said to Villareal, "Now I will crush your skull."

Just then, Petronelli came running out of his office, yelling. He came between us, saying, "I told you not to come here! I called the police and they are on the way. You have to leave."

I said, "OK, Richard everyone here knows what happened." Villareal called me a Filipino wannabe. I told him Filipinos were ashamed of him.

I then challenged Villareal to go with me down onto the median strip on the freeway where everyone could witness our duel, but traffic would prevent anyone from interfering and we could finish it. I'd fight him with his choice of weapons: stick, knife, or stick-and-knife combo. Which would it be? He would not accept. Villareal and Petronelli had just proved they were both "paper tigers". Can you imagine Petronelli and Villareal claiming to be 10th degree grandmasters while along with their friends they outnumbered us five to one, plus he had *called the police*? It was a shameful and cowardly display by any standard.

There was no point to ending up in handcuffs, so we left. Just a few minutes later, once the police had arrived and they were safe, Petronelli called, mouthing off in front of the room full of people and telling me to f*ck off. I quietly told him, "Richard, we both know you're a coward and someday ... it will be just you and me." Dan and Randy just shook their heads at his insolence.

> "The coward only threatens when he is safe." —*Johann Wolfgang von Goethe*

Consequently, the WHC did not approve Villareal's bid for recognition. Subsequently, Petronelli went to Gene and asked him if he would smooth things over. Gene told Petronelli he would not interfere on his behalf.

Fast forward a few years: Petronelli hosted a training seminar featuring ... get this ... Leo. Without so much as a phone call or Facebook message to give me a heads up, Leo was on Petronelli's porch, posting on social media about their seminar.

Since the duel with Villareal, Petronelli has never made any attempt to set things right, so it remains an open wound between us. Richard thought getting publicly recognized by Leo would help him reclaim his honor, but Leo has an infamous reputation for selling ranks and in the Filipino martial arts community, getting ranked by Leo is like thinking you can pick up a piece of poop by its clean end. I stood in for Leo in a duel in which Richard tried to screw us both, so for Leo to train him *for money* was appalling, but not surprising.

I messaged Leo and asked if he remembered that it was Villareal and Petronelli who said "F*CK Leo Gaje". I asked him if he remembered that I risked my life, limb, and liberty in the duel on his behalf. He responded by saying it was a while ago and I should move on. I assured him I had moved on, and as I moved through the last twenty countries, I had remembered to bring my honor with me and I asked him if he could say the same. I told Leo I would drive over to Richard's house right then so we could settle it face-to-face, but he declined. I'm glad I saved the screenshots of that conversation.

That was the last time I spoke to Leo. Leo put his life on the line for my guys and me in the Philippines, and I put my life on the line for his honor in the USA. I repaid his action with loyalty and he betrayed me for a few measly dollars. From Leo, I learned one of the most important lessons:

> *Life Lesson: A man's potential for greatness can slip through the cracks in his integrity.*

Villareal has since passed away, so unfortunately, we'll never get to finish our duel. The matter with Petronelli remains unresolved. And Leo Gaje Jr. has secured his soiled reputation as a greedy prostitute who sold his family's legacy to a clown for a pittance.

Undeniable Truth

> "I went to meet the master but it was his student I was looking for." — *Tom Sotis*

On one of the trips to train in the Philippines, our group was joined by an instructor named David Gould, a student of the very formidable Edgar Sulite (RIP) and other renowned instructors.

On one of the first days of training, I watched Dave working with Leo. It was like putting two glasses of water in the sunlight. Where Leo's movements were cloudy with particulate matter, comparatively, Dave's were crystal clear. I had never before, or ever since, seen anyone move a weapon as effectively as Dave.

I have only ever trained in different fighting methods to become a better fighter, not because I wanted to learn any particular system. I always valued and searched for functionality over form, and in meeting Dave, I felt like a miner who finally strikes the mother lode.

Many people in martial arts place great value in their martial lineage, which is perfectly fine as long as they don't delude themselves or their students into thinking that factor automatically transfers into their own fighting skill. Streetfighting makes one thing abundantly clear: fighting systems and their historical lineages are artificial constructs, a means and record of transmitting knowledge that holds value for its pragmatic use and personal expression, but authenticity truly resides only within you.

> *Life Lesson: Knowledge can be passed, but skill must be acquired.*

I don't at all regret going to the Philippines to meet a truly functional master of the blade, even though that person turned out to be an American. It was a terrific cultural experience, and I found what I was looking for anyway, just not in whom I had anticipated.

The first time you face off with Dave will be forever seared into your memory by the intensity of mentally-blinding pain from an invisible strike. The first thing you see when your eyes and mind can form an image again is Dave, with a "cat-who-ate-the-canary grin", poised to do the same again any instant he desired. When you have spent the better part of twenty-five years training and fighting only to square off with someone who could easily take your life in the blink of an eye, it is truly a very humbling experience.

Life Lesson: You are only as good as you are right now, in this moment.

Combat is an activity of self-evident truth. Unlike conventional sports, there are no points or rules. There is only delivering or receiving pain until you are victorious, or at your opponent's mercy. You must train realistically if you are to have a realistic expectation of your ability to manage all that is involved. A lot of people train hard but not realistically, and for all their effort, still don't fare well under real circumstances. I've had a number of influential teachers, but "Guro Dave" taught me the principles that I had been searching for, and set me on the path to functionality. In my opinion, Guro Dave is the best weapons man on the planet.

Life Lesson: You must begin with true attack, or all learning that follows is flawed.

One Slip

"Rule your mind or it will rule you." — *Buddha*

One of my camps in South Africa was attended by Stan Schmidt, one of the highest ranking black belts in the Japanese Karate Association with many thousands of students under him. Stan was a lifelong martial artist of significant accomplishment who hard-earned the respect he received from so many people.

I always work hands-on with everyone who attends, and I asked Stan if he would assist me in presenting a technique to the group. As I was explaining to the group what I was about to demonstrate and not paying attention to Stan, for reasons I can only speculate, he punched me hard in the face.

Things moved so slowly that time seemed nearly suspended and I could see the shocked expression on the other trainees' faces. When I looked at Stan, he was in a fighting stance with his guard up. I snapped out a backfist and smashed him in the face. Time returned to normal as Stan immediately withdrew, claiming that he was "only kidding". I said, "Of course – that's why you still have teeth."

I was told months later as word spread about what had transpired, that some of the top guys from Stan's organization, along with thousands of its students, resigned.

> *Life Lesson: Just one moment of an unrestrained ego can damage a lifetime of work.*

Judas Is Her Name

> "I have learned to hate all traitors, and there is no disease that I spit on more than treachery." — *Aeschylus*

A couple, originally from South Africa, had trained with me in the USA and moved back to their home country in 1998. Mark Human and Kelee Arrowsmith got married; I attended their wedding, and I taught a couple of seminars while I was there that garnered very little attendance. Apparently, it just took time for word to get out because when I returned a year later to do more seminars, the first one sold out at 120 trainees, and dozens who were turned away drove 550 kilometers to attend my next one.

I went to South Africa every year for a decade. I traveled the width and breadth of the country, training civilians, martial artists, security, law enforcement, and military personnel. One grueling year, I even held sixteen seminars in sixteen days. Mark and Kelee taught AMOK! classes and seminars during the time in between my visits. Eventually, AMOK! had twenty trainers and was the fastest growing combative method in the country.

Things went along fine for a few years, as the company was growing. Early on, I had allowed some of the company trainers to host their own AMOK! Websites, but eventually it led to a fractured presence for our company. Moreover, some of the sites were outdated, so I requested all trainers close their website, and put a notice directing people to the company's main, and now only, website. They all understood why and complied, except Kelee. She said they needed their own site for business, yet the site was badly outdated and business was dropping off because of the economy.

It is important to note that Mark and Kelee both held full-time jobs and taught AMOK! on the side, so it was not their main source of income – it was additional revenue for them from which I never received one dollar. I told Kelee that I had tried to contact the company trainers listed on their website and almost none of them were active, so I requested that she provide me with a contact list so that I could communicate directly with the trainers. Kelee refused. I had thought we were playing in the same sandbox, but she just drew a line in the sand. I knew Kelee well enough to know she wouldn't challenge my authority unless she was ready for fight, so I backed off in order to let her think she had control.

> "Victorious warriors win first then go to war, while defeated warriors go to war first and then seek to win." — *Sun Tzu*

With the help of Andrew Smith, I secured all the trademarks to protect my company and the AMOK! brand, and then we established professional licensing standards to protect the company and our trainers from liability.

While I was doing that, Kelee hoodwinked Andy Elliott, one of the trainers in Australia, to join her in a plan to take over, or carve a chunk out of, the AMOK! brand. She convinced Andy she could help him get a contract with the Australian Federal Police and that they could form their own organization. She would then promote him to the top position in Australia. Andy was weak-minded, with a fragile ego that craved exalted titles, so he fell for it hook, line, and sinker.

Once they had unified, Andy published an article in *Blitz* magazine, knowing that was a violation of our agreement that would get him thrown out of AMOK! He waited for me to chastise him so he would have an excuse to tell his guys that he wanted to leave AMOK! and go with Kelee, who would claim she and Mark were compelled to break away because trainers were asking for their leadership. It was a smart move on her part, but she underestimated me.

Instead, I contacted *Blitz* magazine directly and informed them that they had no permission to print anything regarding AMOK! without my personal written consent. They protested, saying they already had two more articles written and submitted by Andy that they intended to print. I told them that since we had a record of our conversations, I had legal control of my company, and I looked forward to seeing them in court. They never printed Andy's articles. As I had done before with the other websites, I requested all the AMOK! trainers who had an AMOK! Facebook page to close it down in order to centralize our presence on social media. I also informed the trainers that, *IF* insurance was required in the country where they lived, they would have to get liability insurance to operate (otherwise I would be personally carrying their liability).

This is when Kelee decided to make her move. She declared her intention to start her own company and that she intended to continue using the AMOK! brand name. I said they were, of course, welcome to start their own company, but that they could not use the AMOK! brand that I owned and had spent my life building, the company that provided them with a stream of income without them paying any fees or percentage. I had handed them a free stream of income, but that wasn't enough for Kelee; she wanted the AMOK! brand too.

Kelee said I could not stop them and set up an AMOK! South Africa Facebook page. Right away, it was populated by Andy and his students who were announcing their departure from AMOK! to align with "people of integrity". Together, they waged a negative Facebook campaign against me and AMOK!, trying to obfuscate their treachery. Some of them even posted photos of their new AMOK! South Africa tattoos.

After giving them fourteen years of training and support, Mark and Kelee betrayed not only my trust, they betrayed the trust of every member of AMOK! around the world that they previously called brother. Unlike martial politics fueled by ranks and titles (AMOK! has no rank or title for such matters to take hold) Mark and Kelee's motive was simply a greed-inspired, mutinous attempt to steal the power of the AMOK! brand.

Kelee wasted not a single day before contacting an attorney to secure trademark protection for AMOK! South Africa only to learn that I had already secured it. What a blunder! It must have felt like challenging someone to a duel only to have them stab you in the heart before you even have a chance to take your blade out of its sheath.

Exposed and defeated, Mark and Kelee were forced to close the AMOK! South Africa Facebook page and her guys had to get their tattoos covered over. Mark was crushed, so Kelee contacted my attorney to ask if there was a way to work things out, but that was laughable.

In an effort to distract people from their public shame and humiliation, Mark and Kelee attempted to salvage their forever-soiled reputations by trying to damage mine further (as they will try to do again). After having trained the South African Special Forces (SASF) years before, I had their permission to list them as clients on my website, so long as I didn't post photos of any of their soldiers.

It is important to note here that Mark's father was in the South African paratroopers for over thirty years and knew everyone in their special force community. By remarkable coincidence, within a week or two of their failed betrayal, the SASF withdrew their permission to list them on my website and without the courtesy of sending me an email asking me to remove it, posted my name on their Wall of Shame which was then "discovered" and leaked on Facebook to mar my spotless professional reputation. I sent the SASF photos to prove that indeed I had trained the 5th regiment in Phalaborwa and I had kept my word by not putting the photos on my website, but it mattered not, nor did I expect it would.

Considering the present cast of criminals, rapists, thieves, and killers populating the South African government and military, that the SASF would so easily discard their honor was no surprise. Having taught over 1000 seminars in twenty-four countries, it speaks to the high quality of my reputation that the only negative post about me on the Internet involves the SASF. I thus proudly point to my name on their page as a testament to their lack of integrity, and absence of honor.

I genuinely liked Mark. He was a good friend and I always enjoyed his company. The other guys and I always felt a little bad for him because he is the stereotypical hen-pecked husband controlled by an older, domineering wife, but my sympathetic sentiments evaporated into contempt for his lack of a backbone in thwarting his wife's betrayal, and participating in her treachery. After many years of teaching them, they taught me two important things.

> *Mark's lesson: A friend is a friend until the day he isn't your friend anymore.*
>
> *Kelee's Lesson: Some people who pat you on the back are only looking for a soft spot to insert their knife.*

Killers Among Us

> *Hitchhiker: Thanks for picking me up. How do you know I'm not a serial killer?*
>
> *Driver: What are the chances both of us are?*

When I had the investigation agency, I was undercover as a guest in a hotel during a convention. The hotel had its own security staff tasked with preventing drug dealers and escorts from operating. Both dealers and escorts would call rooms and ask anyone who answered if they were looking for anything, so I waited in an empty room for the phone to ring. When it did, the woman on the other end asked if I wanted to date, so I said yes. She asked what room she should come to; I told her I had guests, so I wanted to go to her room and she gave me a room number.

Two hotel security guys and I went down the stairs to the floor below and they waited in the stairwell while I went down the hallway and knocked on the door. The plan was to go in and make the deal with the woman, and then when she took the money, I was going to come and call security so they could escort her out of the hotel.

Instead of a woman named Diane answering the door, it was a man. I asked him: "Where is Diane?"

He said, "She went to the lobby to get some coffee and she'll be back in two minutes. Do you want to come in and wait?"

Security didn't wait for me to go into the room, make a deal, and come back out; for some reason, they had left the stairway and come down the hall. So, as I was talking to the guy in the doorway, they were beginning to approach us. The guy saw them and he looked at me and he asked, "Who are they?" because you couldn't see from the emblem on their blazers that they were security. And I looked at them and said "I don't know. A couple of chumps". The security guys gave us dirty looks and continued to walk away. I guessed they were trying to get a peek at what was going on in the room before I went in, but I didn't know them and that wasn't the plan; nonetheless, their nasty sneer did a good job of keeping my cover.

The guy invited me to the room. I went in, and he closed the door. I thought he was either her pimp, or more likely, this was a set up. I walked into the room, which had two double beds. He asked, "Can I get you anything?"

I said, "No, I'm good." And then he came at me. I caught him with an open hand palm strike to the side of his face/jaw/ear and he sailed over one bed, landing between the two. He was stunned and hurt but still mad. When he got up, he got into a ready stance, so I drove a front kick straight into him. He fell back onto the night table between the beds and then came off it with his guard up, so

I drove another kick into him. This time, I hit him square in the solar plexus and he crumpled.

I opened the door to let the security guys in, and they had abandoned me! I couldn't believe it. I thought that they would've come back after I went inside, and waited right outside the door. While I was doing that, the guy called the desk and said that a madman had attacked him in his room. I went back to him and he stayed on his knees with his hands up, begging me not to hit him again, saying the police were already on the way. His whole body was trembling, not just his hands. He was not just afraid (I'd seen fear enough times to know its face), but was in terror, which seemed odd or somehow out of proportion to the circumstance. But at the time, I just thought he was a coward and chalked it up to that.

The police must have been across the street when they got the call because about a minute later, they were coming down the hall with the two security guards behind them. The first thing they did was grab both of us and put us up against the wall to frisk us. The cop called out GUN! So his partner stepped back, drew his firearm and pointed at me, telling me not to move. The policeman took my firearm and put me in handcuffs. They frisked the other guy and he had no weapons at all. I said to the cops, "I'm here working on behalf of the hotel – just ask security, guys." But they were gone – again! So I explained the entire scenario to the cops. They asked the other guy what his side of the story was and he said he had no idea what was going on. There had been a knock at the door, I had asked him about a woman, and then had come in and attacked him. So the police took his ID, asked him some questions and let him go. The police took me down to the lobby and within a couple of minutes they got the call on their radio that I was a licensed detective, and was licensed to carry a firearm. So they returned my firearm to me and said I could go.

I was angry, so I went back to my room, packed my things, checked out of the hotel, and went back to my place. That night, the Warwick police called to ask me some more questions. It turned out that after getting the guy's ID – while they were taking me down to the lobby – he had packed and slipped out the side door. A while later, a report came back to the police because that man was wanted on a triple homicide in Miami and they were looking for any information they could to try to track him down and find him.

The next day, I was driving down Thayer Street in Providence, Rhode Island and my friend in the passenger side of my car said, "Hey look! That guy sitting in Laguna's outside café looks like he has a purple hand growing out of the side of his head." I looked over and I couldn't believe it - it was *him*! My first instinct was to arrest him myself, and how cool that would be in the newspaper, but I thought it through and decided against it. There were people sitting all around him who would be put at risk and I had no way to know if he had acquired a weapon of any kind. I pulled over and told my friend to keep an eye on him while I went into Andrea's Restaurant where there was a pay phone just inside the door. I dropped dime on him.

Describing him, and his attire, I added, "He's the guy who looks like he has a purple hand growing on the left side of his face." In just a few very short minutes, a couple of plainclothes detectives jumped on him, slamming him to the ground. The café was instantly swarmed by uniformed cops keeping people safe while they handcuffed him and took him away. One of the detectives came over, took my ID, and gave me his card.

> *Life Lesson: You never know if a stranger you're encountering is a murderer ... or perhaps even someone you think you know.*

Quicksand

> "The most important and most difficult thing that you can change is your fixed false beliefs." — *Debasish Mridha*

When I lived in Florida for a short time, my neighbor told me he sometimes went fossil hunting for dinosaur bones in the phosphate mines. I had never heard about that and he asked if I wanted to go.

Jay told me to wear dirt-colored clothing and bring some water. It was only about an hour away to the mines, and on the way, Jay explained that dinosaurs got trapped during extinction and their fossilized bones are mined by cranes that have buckets the size of a small house. He said to look for dark objects that stand out from the sand around them.

When we arrived, we rolled around in the dirt a bit to help camouflage us from the helicopters that spot trespassers. We climbed atop one of the embankments and I could see where the cranes scooped out enormous rows of earth that had to be at least a hundred feet wide and thirty feet down past the water table which formed muddy ponds in each of the rows.

I spotted what looked like a big shark's tooth on the edge of the water and headed down to get it while Jay went down the other side of the embankment. I walked out to the edge of the water and as I bent over and picked up the tooth, the surface cracked much like the way ice breaks under your feet and my foot plunged into the mud. I tried to lift it up which only succeeded in breaking through with the other foot and now both feet were in the mud. It was like falling through the ice in slow motion. Any effort to lift yourself out results in a corresponding downward pressure that sucks you further in.

OK, I was really stuck, so I stood still to think it through and when I noticed I was still sinking, the realization hit me ... I'm in QUICKSAND! Growing up, I had only seen quicksand in Tarzan movies, so unless I was going to an African jungle, I never had any expectation of encountering it. I started calling for Jay, but he was out of earshot. Soon, I was in mud up past my hips and I was starting to get concerned. I kept calling for Jay, wondering where he was. I hoped he hadn't fallen and broken a leg and was calling for me.

By then I was waving my arm, hoping a helicopter would spot me as I'd rather be arrested for trespassing than drown in quicksand. I was thinking how ridiculous it was to be in this situation; when you have time to ponder your impending demise, a lot goes through your mind. Finally, I saw Jay's head pop up over the ridge and I remember the look of shock on his face when he saw how deeply I was buried. He climbed down the embankment, took off his pants, tied a knot at the end of both legs, and came near to the edge, stomping along the way to test where the edge was strong enough. Then he threw me a pant leg and I grabbed the knot with both hands. He sat down, leveraging backward, and eventually pulled me out.

I hope my readers never find themselves in quicksand, but like firewalking, the lesson becomes more valuable than the experience when applied in everyday life.

Life Lesson: Childhood beliefs and assumptions that carry into adulthood can be dangerous things.

Visit to Hell

"Man's inhumanity to man is only surpassed by his cruelty to animals."—*George Bernard Shaw*

Southeast Asia is so different from where I live in New England, it might as well be on another planet. The climate produces a tropical forest with different kinds of trees: palm trees, bamboo, rattan, plus a wide variety of plants, exotic flowers, and colorful vegetation not found in the four-seasons climate of our majestic forests.

The jungles are populated with strange animals, colorful birds, and of course, insects. Big ones. Every walk in the jungle of Southeast Asia feels like a childhood adventure with eyes wide open to the splendor and wonder of the surroundings, experiencing so many different things it seemingly enhances the senses so that one takes in and processes more of the experience.

The roads are similar to ours, but the vehicles can be quite different because less stringent safety regulations permit a wider range of scooters, motorbikes, and other interesting designs of motorized contraptions. In the rural areas, it is not uncommon to see carts pulled by water buffalo on the streets.

One of my favorite places in the world to be is in the area known as "The Golden Triangle" where the borders of Thailand, Laos, and Myanmar meet at the confluence of the Ruak and Mekong Rivers. This far out into the countryside is where I enjoy visiting most because it retains more of their traditional cultures, and is less affected by the modernizations of cities. People who have never been to Southeast Asia find it hard to imagine why some people go there and then choose never to go back from where they came, but having been there, it becomes obvious. I am not saying it is better or superior to other places, I am saying it is an amazing, wonderful, and fantastic place with its own unique beauty and allure that I find very appealing.

Cambodia is a land of stark contrasts, a beautiful culture still recovering from its conflict with the cruel ugliness of humanity. Andrew Smith and Joe Howard came with me to work with the Cambodian Special Forces, Special Police Units, and a private security firm. We were glad to be received by Cambodian officials who escorted us past customs while they loaded our bags into the SUVs and off we went.

The next day, we thoroughly enjoyed a walk through the open-air central market in Phnom Penh. If you want a chicken, pick one out and come back in a few minutes after they've killed and dressed it for you. Fish swim in tubs with a net to scoop out the ones you want; snakes, too. The aisle with bugs has dozens of different kinds in big baskets with plastic bags and scoops. The aliveness of that market is superb, and from this chaos, twice per day there is an amazing display of organized choreography. An approaching train blows its whistle and suddenly in just seconds, dozens of carts and complete food stalls are moved off the tracks (which I didn't even know were there the first time) and the train passes right through the market! It takes roughly a minute for everything to be back again, as though the train never passed through.

We became friendly with our driver and translator, Khun, who shared with us his incredible life story of survival. He was only five years old when the Khmer Rouge massacred his entire village, and Khun was the sole survivor who lived *alone* in the jungle for almost three years before being discovered and brought to a camp. He shared with us vivid accounts and examples of what it was like to be in the camps run by homicidal psychopaths. During that time, an estimated one and half to two million Cambodians died of starvation, execution, disease, or overwork.

Khun took us to visit the infamous Tuol Sleng Torture Prison where thousands of peasants, workers, doctors, students, ministers, monks, diplomats, foreigners, and others were imprisoned, tortured, and exterminated along with their wives and children. Many of the implements of torture remain along with documents, photos, and other evidence. Standing in the rooms where that took place defies accurate description because it affects you on so many levels. Your mind begins to imagine being one of those victims suffering in this hotel of horrors. I've seen my share of violently mutilated bodies, but this level of combined physical agony and psychological anguish was staggering.

After Tuol Sleng, Khun took us to visit the Killing Fields, another monument to the monstrous brutality that we humans inflict upon one-another. As you walk through the Killing Fields you come across a series of stations, each with a sign that describes the way in which people were designated to be killed at that station. The brutality of methods by which they killed the adults was horrifying in its own right, but the children, even babies ....

We exited the fields and came to the Choeung Ek monument with glass walls displaying the skulls of hundreds of victims from which emanates a sense of disquietude. We left and returned to our hotel in silence.

When it got dark, I left the hotel and stood outside by the river in the rain. I thought about the countless similar, unspeakable human tragedies that litter mankind's timeline and I had only questions that drilled down into the deepest recesses of my soul.

What does it take to twist people's minds so they will commit such atrocities on others? What drives people to this level of madness? People who have never been pushed sufficiently for their inner monster to show itself may deny its existence within themselves, but we are all human, and seemingly, there is a potential for a monster to grow inside of any of us.

In the Killing Fields, there is one tree with a sign that describes how that tree was used many hundreds of times to commit atrocities on infants so horrific I couldn't stop the images from invading my mind. It was like their souls were still screaming and mine could hear them. The raindrops were so heavy they felt like bullets falling from the skies.

Life Lesson: Humans now are simply better-dressed savages than they were eons ago.

PART II: MANIFESTO OF INVINCIBILITY

MANIFESTO OF INVINCIBILITY

VALUES

MOTIVATION

"We value what we desire and we desire what we value." — *Dr. Steven Reiss*

Before exploring values, it is beneficial to understand how our deepest motivations drive them. Just as our DNA determines our unique physical characteristics, so does it also generate our unique individual personalities. Understanding our genetic motivations and how their expression is affected by our culture and upbringing helps us to identify the underpinnings of our individual personal values. Motivation analysts reject the common notion that motivation is about psychological energy and show that motivation is generated by the assertion of one's deepest held values.

All human beings share the same basic sixteen motives that drive every decision and action we make but each person prioritizes and emphasizes them uniquely, which shapes our individual personality and which drives our behavior.

Struggles with self-motivation and self-discipline come largely from not knowing how our sixteen basic motives are arranged, how they operate, and how they subconsciously drive us to behave in ways in which we may not understand, or can't seem to control.

The profound insight gained through Reiss Profiling increases your control over your inner drives, intensifies the motivations you want to encourage, and reduces your internal struggles with hard-to-control behaviors. When you understand how your genetically-inspired desires are arranged, how they drive your behavior, and how they were shaped by culture and your upbringing, you make better choices that bring about better personal outcomes and improved relationships.

As a tool for achieving peak performance and invincibility, motivational profiling holds profound implications. In any given situation, your Values (desires) dictate your Strategy (objective), which governs your Tactics (plan) that results in your Techniques (actions).

Any flaw between the execution of a technique and your intention to execute it is a vulnerability. Any discrepancy between a tactic and your strategy will achieve an unintended result. Any doubt about your strategy arising from conflicts with your Values will create hesitation and a lack of commitment. If there exists any struggle between your Values and your Motives, it will sap your willpower. Tactical wisdom is that "good technique won't compensate for lack of willpower."

Therefore, the state of invincibility is only possible when you first clearly understand your motives, uphold your values, commit to your objectives, hold true intentions, and execute actions with your entire self.

As a performance optimization coach, I was excited to discover that there is finally a science to motivation, and because lives depend on the quality of my work, when I learned how the Reiss Motivation Profile® (RMP) could be used as a coaching tool, I was compelled to become a certified Reiss Profile Master. As a tactician, I marvel at the genius of Dr. Reiss' mapping of the tactics of behavior. As a coach, I don't struggle to get inside my clients' heads because now I can see right into their hearts.

Developed by the brilliant Dr. Steven Reiss (1947-2016), the Reiss Motivation Profile® is the first scientific test that provides a concrete, detailed image of a person's individual basic desires, motives, and values that is as unique as your fingerprints. It reveals profound insights into your genetically-driven motives showing precisely, as an individual, what makes you tick.

VALUES

"Our value is the sum of our values." – *Joe Batten*

With an understanding that our deepest motives are genetically-driven and affected by our culture and upbringing, it becomes easier to see how we, as individuals, pursue and develop our own personal values that shape our strategies.

Personal values reflect your ideas on a range of topics, so I have narrowed our exploration with those values that are primary to achieving invincibility.

If a warrior seeks *invincibility*, the person within seeks *impeccability*. This does not imply that they have no faults; it means others cannot find fault with their behavior. To that end, I strive to be a better man, to maintain honorable intentions, earn more loyalty, show more respect, guard against betrayal, exact just revenge, use failure as my stepping-stone, defeat violence with force, to be kind whenever possible, and help deniers discover the empowerment and freedom that comes with skill.

Brotherhood

"In AMOK!, brother/sisterhood is a *verb!*" —*Tom Sotis*

What motivates a person drives their underlying values about training. People seeking validation through recognition will seek instruction that places value on artificial awards. The extrinsic rewarding of rank and title fosters competition between participants and establishes the footholds for martial politics, ultimately eclipsing the importance of skill development.

People seeking confidence through skill will seek instruction that places value on functionality. Thus, I removed all extrinsic rewards of belts, ranks, and titles, in favor of the intrinsic rewards of skill and brotherhood, the development of which are mutually-dependent and fosters cooperation.

If I trust your heart (brotherhood) but not your physical control (skill), then there are some dangerous developmental practices we cannot do because it is likely you'll injure me, even though you would feel bad about it.

If I don't trust your heart (brotherhood) even if you have physical control (skill), then there are some dangerous practices we cannot do because I cannot trust that you won't become reckless with my safety, and it is likely you'll injure me.

And when I trust your heart (brotherhood) and your physical control (skill), then there are a number of increasingly dangerous practices we can do because it is unlikely you'll injure me, and even if so, I'd know it wasn't due to recklessness.

A brotherhood forms when a group of people act continuously for the benefit of each other. Amoks are drawn together by common values and they bond through shared experiences in training and through mutual support in life. Sharing the common value of wanting to have the skill to match our will in protecting ourselves and loved ones is fundamental to people from every culture on earth, and as such, is a unifying activity that dissolves other social constructs that might otherwise divide us.

> "Fascism is cured by reading and racism is cured by traveling." – *Miguel de Unamuno*

Arguably, what one reads and to where one travels could exacerbate fascist and racist ideologies but, taken in a context that assumes one becomes widely-read and widely-traveled, thus exposing one to a varied selection of ideologies and cultures, both fascism and racism dissolve when the truth of personal experience differs from previously-held assumptions and limiting beliefs.

Fighting has the same "leveling effect". If you travel widely and fight a wide spectrum of fighters, you are certain to be humbled by the experience. When you fight, you learn about yourself because fighting tests your beliefs and values about yourself, and at times, your world view.

A few years ago, after a long day of training at our Thailand camp, I looked around the dinner table and took stock of who we were: different races (black, white, Asian, Latino, mixed); different genders (straight and LGBTQ spectrum); different beliefs (Christian, Muslim, Jew, Baptist, Buddhist, atheist); different politics (conservatives, liberals, socialists, anarchists, etc.); different food preferences (carnivores, vegetarians, vegans), and we came from different occupations from all walks of life. There we were – this wonderful group of widely diverse people – eating, drinking, sharing stories, and laughing together. Dining with these fine folks, I felt a great sense of satisfaction and contentment.

Honor

> "You can abandon your own body but never let go of your honor." – *Miyamoto Musashi*

Honor is often understood in vague terms surrounding someone's ideals of taking noble actions on behalf of someone or something. Different cultures develop varying ideas as to what constitutes honor, and what distinguishes matters of honor from other matters. In some cultures, matters of honor and reputation are a distinction without a difference. Conflating honor with reputation leads to perversions of honor in which crimes and atrocities are committed in its name, but are, essentially, little more than horrific expressions of bruised egos wrapped up in cultural excuses.

> "Rationalization is used to circumvent the standards of your principles but the truth cannot be hidden anywhere within one's own heart"- *Tom Sotis*

In his book, *Living the Martial Way,* Forrest E. Morgan explains what are the essential elements of honor, the different levels of keeping honor, and what separates matters of honor from matters of reputation better than any other source I have encountered. Below, I have encapsulated the framework Mr. Morgan presents in the three required elements and the three levels:

The Three Required Elements of Honor are moral obligation, justice, and courage.

- Moral Obligation - The first required element is that you must have a moral obligation to act. This means that either you are a sworn officer, soldier, or agent compelled to act, or in civilian contexts, your personal morals compel you to act (i.e., you see a child about to be abused).

- Justice - The second requirement is that your action must be a just action with no collateral damage. This means if someone hurt your loved one, you cannot in turn hurt their loved one for revenge because it is unjust to that person; nor can you permit anyone else to be incidentally harmed when exacting revenge on the subject.

- Courage - The third requirement is that once you have the combined elements of a moral obligation to act and a just action to take, then you must take that action, even when it requires personal sacrifice (putting principle over consequence).

The Three Levels of Keeping Honor posits that none of us are pillars of perfect morality where errant thoughts don't occasionally cross our minds. Mr. Morgan teaches us that the virtue in honor is not just in keeping it, but in the *how and why* it is kept.

- Highest - It never crosses your mind to do the wrong thing before doing the right thing (Right action, no reasoning occurs).

- Middle - It crosses your mind that you *could* do the wrong thing, but you immediately dismiss it and do the right thing. (Right action, right reason).

- Lowest - You first decide you *will* do the wrong thing and then afterwards, decide do the right thing only for fear of being exposed. (Right action, wrong reason).

The three requirements are vital in separating matters of honor from matters of saving face. In some cultures, and with some individuals, when someone feels another has marred their reputation or committed an offense that causes embarrassment, the perceived victim may feel entitled and/or compelled to resort to violence in order to regain their honor. Indeed, there are a number of reasons why your reputation may be worth fighting for, but they are not matters of honor when there is no moral obligation to act. When there is no moral basis for violence, rationalization is a tool people use to fabricate moral justifications for the immoral acts they commit.

"To engage in rationalization is to tempt one's honor." – *Tom Sotis*

The seductive power of rationalization is a slippery slope, like a moral "get-out-of-jail-free card" bound only by one's own conscience, or lack thereof.

I think it important to recognize the relationship between honor and rationalization because they form the arena where we fight to preserve our integrity. In guarding our integrity, I have identified Three Levels of Rationalization to which we must remain vigilant.

- Reason - make a good case that the ends justified the means (wrong action for a right reason)
- Excuse - make a poor case that the ends justified the means (wrong action for dubious reason)
- Immoral - action has no reason or rationalization (sociopath)

"Rationalization puts a pebble in the shoe of your character." – *Tom Sotis*

So, what do morals have to do with Invincibility? Internally speaking, the elements of morals, ethics, strategy, and mindset are what drive your actions, and if they are not properly geared and engaged, performance suffers.

Rationalization (the action of attempting to explain or justify behavior, or an attitude with logical reasons, even if these are not appropriate) occurs because you desire to behave in a way that conflicts with your ideals. This is the root of all doubts (a feeling of uncertainty or lack of conviction). This is where your conscience struggles for truth; it is the place we commonly describe as "deep in your heart".

When your conscience is not absolutely clear – if you have any doubts or reservations about your strategy or plan – your actions will suffer in their execution.

Loyalty

> "You cannot buy loyalty; you cannot buy the devotion of hearts, minds, and souls. You have to earn these things." — *Clarence Francis*

The noble word *loyalty* is often subverted to emotionally bully people into going along, or to take advantage of someone's trustworthy and devoted good nature.

There is but one way to earn honor: When someone makes an *extended or significant personal sacrifice* on behalf of another person *of good character*, that person will naturally feel a sense of personal obligation to their benefactor.

If there was no extended or significant personal sacrifice, then loyalty was not earned; perhaps a favor owed (honor-bound), but not quite one's loyalty. This is why it is prudent to guard against becoming indebted to someone who may ask you to compromise your honor in order to settle said obligation.

If the person for whom an extended sacrifice was made was not of good character, and the sacrifice was not appreciated, they will not feel any sense of personal obligation (i.e., he or she will be ungrateful) and as such, will offer no loyalty.

If the person who has earned loyalty makes even one action that is not in the best interest of the other, they may potentially lose every loyalty credit ever earned, especially if it is compounded by a breach of trust.

People of good character who receive an extended or significant personal sacrifice are grateful, while those with poor character are not only ungrateful, but often resentful, and this is the root of betrayal.

"No good deed goes unpunished." – *Unknown*

This is why people of good character who help others are surprised when people they helped talk badly about them behind their back, and sometimes even seek to actively betray them. Honorable people can't wrap their minds around resenting someone who helped them – their moral compass doesn't point in that direction, so they get burned repeatedly until experience produces more caution. Betrayal can only come from someone who was previously trusted.

Respect

"Respect everyone." – *Traditional wisdom*

Showing respect is acting in a way that conveys a sense of admiration for good or valuable qualities. Generally, people show respect for those whom they admire, and do not show respect for those whom they loathe. There is a difference between respecting someone, and showing them respect.

Tactical wisdom admonishes us to respect everyone; where "re" means *again*, and "spec" means *to look* (spectacles), so at its root, the word "respect" means "to look again" and implies that one is well-advised to look beyond the obvious, to look below the surface, to see the nature of a thing apart from, and to not be deceived by, its outward appearance.

The English idiom, "Don't judge a book by its cover" applies quite particularly to fighters. Some of the most formidable warriors come in the most unsuspecting human packaging. I have come to know hundreds, perhaps even thousands, of average, everyday folks who although they may appear to be perfectly harmless or unskilled at fighting, they are actually highly skilled, and capable of extreme personal violence if forced to defend themselves or their loved ones. This has been especially true of a number of people I've met and worked with in the intelligence community. Some of the guys I've trained ... At first glance, one might think they are just guys in suits, but they could take your life in a heartbeat.

Oftentimes, after meeting me in social settings, when some people learn from others what I do and who I train for a living, they say, "Really? That guy?" with some level of doubt or disbelief, which makes me pleased. I'd rather that people first see my other qualities before they start making assumptions about what kind of person I am based on my particular skill set.

I've had an adventurous life, but not an easy one, and I have landed in situations far from home with no place to stay and nobody to turn to, so I know what it is to live on the street. I have learned some interesting and important things by sharing meals with homeless people. When you respect them enough to share a meal, as opposed to just buying them one, and simply ask "How was your day?", people will open up to you. Everyone has a story.

Betrayal

> "For there to be betrayal, there would have to have been trust first." — *Suzanne Collins, The Hunger Games*

We know that when it comes to honor, we humans are principled to different degrees about different things. When I was younger, I tended to be somewhat trusting unless a person had given me cause not to trust them. My optimism in human character often obscured the tell-tale warning signs of potential betrayal, and it took me too many years to learn the following:

> *Life Lesson: Just because you are friendly with someone doesn't mean you are friends.*

Don't confuse likable people in your social circle with those who are truly concerned about your welfare and well-being. After being betrayed, or becoming the victim of some other heinous act, people often say, "Well, he *seemed* like such a nice guy." We must learn to see the nature of a person sans the rose-colored glasses of friendship. When we like someone, we tend to be blind, or turn a blind eye to their character flaws; we may also see those flaws but our ego fools us into thinking that particular flaw will never cast its shadow upon us.

In preventing criminal attacks, we learn that we cannot best judge if someone is a threat simply from their apparel, but more so by their behavior, which gives us indications of their intentions. We stand a significantly better chance of avoiding or escaping from attackers when we know what to look for – and the same thing is true for sniffing out betrayal.

It is helpful to develop a moral profile of people around you by observing their principles and rationalizations. What principles do they adhere to, compromise, abandon, or violate? When and why do they rationalize away actions inconsistent with principles they claim to hold (hypocrisy)? What are the warning signs? Here are some questions to consider when developing a moral profile of someone:

- Do they believe in earning money legally? Or do they rationalize why illegal activity is acceptable (think they are above or outside the law)?

- Do they resent people with more money or net worth, and rationalize stealing from them (they are envious)?

- Do they resent people with more success and claim that successful people have "cheated" the system in some way, or are otherwise undeserving (they are envious)?

- Do they resent people who get more attention (others who are more popular) or those who receive awards or recognition (again become envious)?

- Are they unable to keep secrets? Do they rationalize away their integrity and treat confidential matters as currency (they are untrustworthy)?

- Do they lie to others because it is expedient, and do they rationalize avoiding the truth for inconsequential matters and will lie when it is a matter of consequence for them (they are untrustworthy)?

- Do they use people? And in so doing, do they rationalize that the person somehow owes or is obligated to them (they feel entitled)?

- Do they lack empathy, professing that it's a dog-eat-dog world and victims deserve their fate (they identify with wrong-doers)?

It is important to recognize patterns of behavior in terms of which *principles* a person adheres to, which ones they don't, and the types of *rationalizations* they use. All of this combines to construct a profile which reveals his or her particular character strengths and weakness, and thus, should help to predict the ways in which they may be trustworthy, and ways in which they may not.

Revenge

> "Revenge is an act of passion; vengeance of justice. Injuries are revenged; crimes are avenged." — *Samuel Johnson*

Matters of honor are always a matter of justice, but matters of justice are not always a matter of honor, but of justice itself. When legal remedies fail to serve justice, it is possible to exact revenge and keep your honor, but your retaliation is subject to three conditions:

1) It must be proportional to the damage incurred or it creates an unjust imbalance, initiating a cycle of violence.

2) It must commit no collateral damage whereby no uninvolved people are negatively affected, or it creates an injustice.

3) The target of your retaliation must know it was you who retaliated. If taking open revenge against an adversary with

no honor would endanger your loved ones, then and only then, is taking secret revenge considered acceptable.

Rightly or wrongly, someone could rationalize an act of revenge against you aimed at causing you various types and degrees of harm. That spectrum includes, but is not limited to: frustration, anger, property loss, fear, sorrow, humiliation, isolation, pain, or death.

To help us identify and remain vigilant to the many ways people might try to take revenge for real or imagined wrongdoings, below is an *incomplete* list of ways in which people can exact revenge:

1. Electronic harassment through phone calls, emails, texts, computer hacking, etc.

2. Interrupting logistics/operations by sending/intercepting deliveries, sign-up/canceling of services

3. Initiate governmental pressure through false reports that instigate inquiries, inspections, audits, and investigations.

4. Damage to possessions such as a home, car, boat, motorcycle, bike, or anything else unprotected.

5. Attack someone's finances by trying to get them fired from work, create credit problems, create/report conditions that result in fines, or by stealing their assets.

6. Restrict their freedom through police reports, restraining orders, getting you arrested, or even imprisoned.

7. Create confusion through misinformation; mislead, misdirect, cause vital items to be misplaced, steal items, deleting computer files, etc.

8. Attack their reputation through rumors and social media.

9. Isolate someone from the support of their family, friends, clubs, church, any other groups to which they belong.

10. Attack their health by placing dirty needles for them to accidentally get pricked on, run their toothbrush over a heavily-saturated area of bacteria, or otherwise contaminate their food/beverages, etc.

11. Cause emotional pain through jealousy, anger, fear, sorrow, guilt, catch-22, etc.

12. Expose secrets with regard to sexuality, crimes, lovers, relationships, children, money, work/personal history – anything they don't want others to know.

13. Threaten loved ones, whether it be their children, spouse, relatives, friends, or even their pets.

14. Cause physical pain through beating, fire-burning, chemical-burning, cutting, drowning, crushing, suffocating, choking, injecting, inserting, etc.

15. Arrange certain death, where someone controls the time, the place, and the manner in which the target of the revenge will perish.

This is why I repeat a quote from my earlier lesson of the *Lone Wolf* story...

> "One enemy can do you more harm than 100 friends will do you good." — *Anonymous*

Failure

> "Failure is a currency by which we purchase experience." – *Tom Sotis*

If experience is the best teacher, then it is through failure that the most valuable experience is gained. Typically, and unfortunately, there is a directly proportional relationship between failure and experience; the greater the failure, the more valuable the experience.

Whenever you get hurt in a fight, the very first thing you do when you get back to training is develop a solution to prevent or counter that hurtful action. Typically, one's commitment to developing that solution is directly proportional to the amount of pain the failure cost them. Modeling that, the AMOK! method is to fight first in order to expose a personal weakness.

1) Identify the weakness

2) Discover a Solution

3) Practice the Solution, and

4) Train yourself. Repeat.

AMOK! is the process of identifying weakness through failure, and then repairing them, but not just the physical weaknesses. Indeed, *Invincibility* is developed by climbing the internal steps of failure too. Each one of those steps has a name: pain, ego, disappointment, loneliness, isolation, disillusionment, betrayal, doubt, etc.

If we value functionality and embrace the valuable gains that failure has to offer, then when we engage with others, we either win or learn, but we don't lose.

- Winners, if they're smart, learn a "soft lesson" by recognizing what they did well and how they could do better.

- Learners, on the other hand, may lose an engagement, but learn a "hard lesson" that they will address so as to never repeat it.

- Losers make excuses and/or blame others, denying their lack of performance or ability and learning nothing, which is why they are losers.

Violence

> "Though defensive violence will always be 'a sad necessity' in the eyes of men of principle, it would be still more unfortunate if wrongdoers should dominate just men." - *Saint Augustine*

Dr. Robert Sapolsky, perhaps the world's leading neurobiologist, gives a brilliant speech on TED Talks entitled, "The Biology of Our Best and Worst Selves", in which he traces the factors that influence a violent behavior. As Dr. Sapolsky points out, as humans, we are not against violence, we are against *certain types of violence in certain contexts.* Personally, I despise violent people who victimize innocent people and have dedicated my life to helping others protect themselves from them. Bullies and thugs beat people up. Some thugs cross the line and become rapists and killers, while psychopaths enjoy torturing.

The first Law of Nature is self-preservation. All living creatures struggle to thrive and survive in their place along the food chain. They will kill in order to eat and not be eaten. Similarly, if you attempt to hurt me, I will protect myself. To *practice* natural law, humans admonish "The Golden Rule: Treat others as you wish to be treated." To *enforce* natural law, humans developed "The Justice Rule: Use necessary force to stop unjust violence."

> My son, Jake: Dad, what's the Golden Rule?
> Me: Treat others the way you want to be treated.
> Jake: But what about bad guys, Dad?
> Me: Well, son, we have *AMOK!* for them.

Humans have used tools to commit acts of violence and have used force against each other since the dawn of our species. In fact, the evolution of tools used as weapons has been a predominant factor in the development of our species, cultures, and history. First with rocks and sticks, then slings and bows, and eventually evolving into swords and guns: while the technological means of violence and force have evolved tremendously, biologically-speaking, humans are still the same creature. The fundamental right of being human is self-preservation against threats to our survival, which naturally includes having the means *equal to the threats.*

I was the first person in the USA who dedicated his life's work to normalizing the use of knives for self-protection. Of course, other guys before me taught a little knifework, but nobody had the vision, or commitment to the vision, to bring knives to the self-protection industry as its own category. I thought, *we have martial arts and we have firearms, why don't we use knives too? Why should only criminals enjoy the benefits of that tool?*

Initially, people thought I was an extremist, but twenty-five years on, the self-protection industry has caught up as tactical folders are sold by the millions, and there are now hundreds of knife instructors. Unfortunately, all but a few of these instructors are, as we call them, "virgins teaching sex", and we know it as soon as they begin to speak about it, but to each other and their students, it all sounds perfectly plausible. Even when they are well-intentioned, lives can be put at stake by an inexperienced instructor. They present a very real danger and in my professional opinion, in most cases, having no training is better than receiving the wrong training.

I wish my profession didn't exist, but it will for as long as we produce violent people. Sometimes I wonder if the human species will survive long enough to evolve to the point where we look upon violent confrontations in the same way we now view and are revolted by cannibalism.

STRATEGY

OUTCOMES

> "Without strategy, execution is aimless. Without execution, strategy is useless." – *Morris Chang*

Strategy is all about outcomes. What is your objective? What do you want to accomplish? Oftentimes, one's strategy is clear, but other times, it can be vague and problematic. In some instances, a strategy can be very clear and still not generate a Best Possible Outcome (BPO) because it was off the mark.

In order to achieve BPOs we must:

- Pre-examine our motivations by thinking things through to their logical end (beyond the immediate), assuming all facts will be publicly known and you will be held accountable for your actions
- Establish strategic protocols for different types of situations
- Make necessary preparations

In matters of self-protection engagements, there are four likely outcomes where, afterwards, you either go to the morgue, the hospital, the jail, or home.

To avoid the morgue or hospital, you must be *tactically-sound,* whereby your actions must resolve the immediate threat. As well, your threat-resolving actions must be *legally-defensible* if you are to avoid going to jail/prison.

With these two vital elements satisfied, ideally, your actions will also be *socially-responsible* so as not to damage your reputation and invite negative consequences into your business or personal life. And finally, your actions should be *morally-*

acceptable (according to your own moral compass) such that you do not act in ways that could result in lifelong regret.

ASSESSMENT

> "It is just good practice to do a situational scan and have situational awareness when you are out in the world."- *Phillip C. McGraw*

In furtherance of Best Possible Outcomes, it becomes necessary to create an accurate situational assessment of potential danger and threats so that we can identify, prioritize, and emphasize safety measures to mitigate them.

Situational awareness recognizes "anything could happen anywhere" and still, context is king. Some places are inherently safer than others where you may not even make a mental note with regard to your safety, while other places are inherently dangerous, requiring more of your dedicated attention to be adequately aware of happenings around you. This permits a natural adjustment to one's environment (so that you won't burn out being overly concerned where there needn't be any because that's not relaxing), nor do you want to let your guard down because you were inattentive at the wrong time (which is not prudent).

Danger and threats can appear in many places in many forms, so it can be beneficial to have a simple mental protocol that follows a natural progression of thought that doesn't have to be memorized or mentally check-listed, especially under stress.

Consciously practicing this protocol in everyday situations trains your mind to habitually scan your situation and accurately size it up. Soon, you won't need to consciously scan because you will begin to perceive your environment through these "safety filters", and with no special effort or attention, you will enjoy a significantly elevated ability to detect, avoid, escape, or respond to moments of crisis.

These four questions size up your situation:

- Environment - what sorts of dangers are presented by my environment?
- Inhabitants - which people present a potential threat?
- Behaviors - who is acting in suspicious or questionable ways?
- Instincts - what are my instincts telling me?

These four questions size up your "just in case" protocol:

- Escapes - where are the exits?
- Improvise - what items can I use to escape?
- Diversion - what can I do to aid in escape?
- Communication - how can I lead others to safety?

STRATEGIZING

> "In preparing for battle I have always found that plans are useless, but planning is indispensable." — *Dwight D. Eisenhower*

When we are under time-sensitive pressure, our brain's neo-cortex (problem-solving apparatus) passes major functioning control to the reptilian parts of our brain (survival functioning) and as such, we lack the momentary capacity to form logical assessments and strategies.

We all know what it is like to make a quick decision under pressure and then afterwards kick ourselves, asking: why didn't we do this or that instead? This is why hindsight always has 20-20 vision. We are all familiar with the adage: Good judgment comes from experience, which comes from poor judgment. *Hindsight.* And to that I would add: Good judgment comes from thinking it through in advance, before the problem presents itself. *Foresight.*

For this reason, it benefits us to pre-examine our will, and to develop response strategies for potential threats so that if we should ever be presented with a similar situation, we will have already *thought it through to a logical desirable conclusion in advance* and we can confidently take actions towards that end. Pre-examining our will (strategizing) helps us to develop foresight, which may not have the clarity of 20-20 hindsight, but is far better than entering into situations blind and unprepared.

Another benefit to strategizing is that having given serious consideration to mitigating potential threats, we become better at the early detection of their presence, which improves our ability to effectively respond. Furthermore, in moments of crisis or peril, pre-examining our will lessens the potential for hesitation because you won't have to take the time to decide what to do since you've already decided, so you can set about acting right away.

Most people practice what I call "Walter Mitty Strategizing" whereby they imagine some type of self-defense scenario in which they rescue someone from a criminal or predator and emerge the skillful victor, a hero of their imagination. But in real life, Walter Mitty was a pushover.

In order to achieve BPO's we must think things through to their logical end (beyond the immediate) assuming all facts will be publicly known and you will be held accountable for your actions. One way to do that is to imagine yourself standing before a judge and saying, "Your honor, given the same facts and circumstances, I would take the very same action again." After you consider scenarios, solution options, contingencies, and consequences, what are you willing to do?

Mentality

> "My mentality is that of a samurai. I would rather commit seppuku than fail." – *Elon Musk*

A lot is revealed in how a person deals with people and problems. When presented with a problem, I recognize three main factors that converge and galvanize into forming one's mentality. Mentality is the sum total of your objective, your values, and your willpower.

Your objective (desire) subjects each solution to your values (moral standards). Either you select the best solution, or rationalize a new standard (a test of your integrity) until you have one. Willpower is measured by the strength of your commitment to seeing it through (conditional or absolute).

"There is no genius without a touch of madness." – *Seneca*

Mentality is not measured just in terms of commitment, but also in terms of personal expression. It is one thing to be committed to using a knife to protect your life in personal combat – it is quite another to elevate the technical efficiency of doing so to a level that creates its own form of artistic expression.

Knife combat is undeniably horrific and it seems contradictory, if not ironic, that it exhibits its own form of creative artistic beauty, a form of appreciation and respect that, perhaps, can only be observed from a purely tactical perspective.

My public demonstrations normally deliver a sanitized expression that still tends to leave most people shocked, if not aghast, at what someone skilled with a blade can really do. Afterwards, people either want to talk with me, or get as far away from me as possible. Trust me when I tell you that few people want to see what skilled knifers can really do.

Legal Use of Force

> "The intention behind the protective use of force is to prevent injury, never to punish or to cause individuals to suffer, repent, or change." – *Marshall B. Rosenberg*

What constitutes the legal and just use of force against illegal and unjust violence varies widely. There are a multitude of relevant factors surrounding a legally-defensible use of force that are outside the scope of this chapter.

In general, Use of Force matters generally revolve around three main elements of each participant's weapon, intent, and capability.

- Weapon - which weapon(s) did they possess? Was it legal for them to possess the weapon, and did they use it?
- Intent - was it a matter of one person attacking another who is forced to defend themselves, or do they share mutual culpability, and even if they acted in defense? Were their actions intentional or accidental?
- Capability - two people of disparate physical capabilities will be judged differently than two people of equal capability because they are not on a level playing field.

I advise people to find out exactly what the laws are pertaining to self-defense where you live, or wherever it is you are traveling, and then *ABL*, which means Always Be Legal. Do not carry any forbidden or restricted weapons. Even if you have the legal right to self-defense, you won't achieve a BPO if you are arrested for using an unlawful weapon. If you are somewhere in which you feel the need to possess an unlawful weapon for self-protection outweighs the risk of having it, that is a decision you have to weigh for yourself.

"Buying a guitar don't make you a musician" and likewise, self-protection items don't come with the skill to effectively apply them under stress. Get trained in the use of various self-protection devices and weapons. With each one, learn how it works, what it does to people, and how to use it just a little, or a lot. That way, you can use *just the right amount of the right tool for the right problem*.

Tactical Logic

> "In design, be logical, search for truth, be clear." – *Massimo Vignelli*

Since the very beginning of my mission, I've shared what I titled, "The Undeniable Logic of the Blade", which identifies the types of threats we are likely to encounter and then compares the different means of force available (empty-hand, stick, knife, firearm) to achieve best possible outcomes (tactically-sound, legally-defensible, and socially responsible solutions).

Here it is, as concisely as possible:

The spectrum of threats to civilians ranges from the Untrained, Trained/Experienced, a Goliath, Gang, Stick, Knife, Pistol, or Multiple Armed Opponents. In terms of achieving not only tactical victory, but a legally-defensible one, here are some guidelines as to how best to prepare...

- Empty-hands martial arts - longest time to functionality. Gives an inherent advantage over the Untrained, but leaves one vulnerable to the other threats.

- Firearms - long time to functionality. Gives no inherent advantage against other firearms, and some advantage over other threats at a distance but leaves one vulnerable to the other threats in close quarters

- Sticks - moderate time to functionality; offers escalation of force across all categories. Only at a disadvantage against firearms at a distance. Difficult to carry/draw/deploy/employ.

- Knives - shortest time to functionality; offers escalation of force across all categories. Only at a disadvantage against firearms at a distance. Easy to carry/draw/deploy/employ.

When examined logically, not only do knives offer the fastest, most practical, and most effective means across the widest spectrum of threats, but another great advantage is that you can substitute any other hand-held instrument and use the same blade-based movements to fight with or against any other weapons, so you don't have to learn new moves every time you acquire a new weapon.

That said, empty hands, firearms, sticks, and knives are not in competition in any way – they are complementary. We need to use the right tool, the right way, with the right amount of force to get the best outcomes. And since we want to become proficient with all of them, it makes logical sense to start with the one system that lays a foundation for all the others; that way, you effectively establish a "common denominator of fighting".

PROTOCOLS

> "Strategy compels tactics; protocols dictate them." – *Tom Sotis*

Strategies are your desired objectives, and *protocols are established courses of action to be followed* to meet them. Protocols protect you in two very vital ways:

First, protocols provide a best course of action under time-pressured decisions. Under duress, your prefrontal cortex – the brain's problem-solving mechanism – disconnects, thus inhibiting your ability to make best judgments and decisions. This natural defense mechanism exists so that you can't inhibit your fight-or-flight survival mode. After a violent event is over and the prefrontal cortex has kicked back in, this explains why more obvious and better solutions become apparent. It is also why hindsight is 20-20 after an event, and it is common to think: "I should have done this or that instead of…."

Second, protocols guard against emotions affecting critical decision-making, or leveraging alternative outcomes. Our judgment could be unduly swayed by emotional pressure (i.e., what people will do to protect their loved ones). Protocols help to resist temptations in moments of weakness (i.e., when involved in a long-term relationship, you resist advances from someone else you desire because it is appropriate for you to do so). Protocols exist to provide a decision when faced with a moral dilemma (i.e., risking the lives of one hundred people in order to save one).

Context determines whether it is best to establish and employ "loose protocols" that permit a certain degree of flexibility in achieving outcomes (i.e., how a team might accomplish a goal) or "strict protocols" that permit little or no flexibility (i.e., a surgical procedure).

Sometimes, a tactical solution seems simple and straightforward, while at other times it is not. Protocols can develop slowly over time as they are gained by experience, and/or they can be rapidly acquired through realistic scenario training.

After engaging in scenario training, and experiencing how things can go wildly outside of your expectations, one develops more appreciation for protocols and comprehensive approaches to the same situations.

For example: You are in a parking lot and come across a couple arguing. Suddenly, the man shoves the woman against the car door and starts slapping her while she cries out for help. Nearly every guy in training jumped right to her rescue (their natural immediate instinct was to attack him) yelling at him and pushing him off and/or even striking him until he backed off. At this point, the trainee is between the man and the woman, so the woman draws a training knife and stabs him in the back. The couple continue the assault, robbing him of his phone and money; he lays there, bleeding out.

> *Lesson Learned:* A victim could be faking, or may turn on you.

After a few failures, the trainee learns to stay at one end of the car (witness), take and send a photo of the aggressor, and call 911 (report). Then the trainee informs the couple that he has called the police and sent their photo (verbal intervention). The trainee then instructs the woman to come out from between the cars (passive intervention), and assuming he lets her and she does, to stand a few feet away (triangulate them).

If the trainee lures the man out instead, at least he has positioned the attacker between himself and the woman so the trainee doesn't have his back to her while dealing with him. If he ignores the trainee and doesn't stop, the trainee has to decide how to physically intervene. The trainee learns to use grabs to pull him off (defend) but keep the guy between himself and the woman. The trainee learns to use a combination of strikes (attack) and grabs to pull him off, but keeps the guy between himself and the woman.

PREPARATION

"Achilles should have worn better boots." – *Tom Sotis*

Having developed strategies and protocols for various situations, we must make preparations to implement them. The type and scope of preparations is dictated by the event, whether it be combat, self-protection, business, athletics, lifestyle, or whatever goal you are pursuing.

Preparations might include some, all, or more than these areas: personnel, finances, tools, weapons, equipment, supplies, first aid kit, logistics/transport, resources/lawyer, medical, and assets. For most situations it is sufficient to answer the basic 5W's and 2H's: who, what, when, where, why, how, and how much?

Most people think that martial arts and self-protection are synonymous, a distinction without a difference. With regard to self-protection, the vast majority of martial arts have serious design flaws in achieving best possible outcomes because they focus mainly on developing reactive defensive techniques. Below is a list of the major flaws I have observed in conventional martial arts training:

1) Beyond general advice and admonitions, there is little if any actual training or curriculum for all that leads up to the point of first physical contact (lacking proactive measures, just reactive techniques).

2) Questionably-effective techniques address a narrow spectrum of threats (only provide inherent advantage over inept attackers).

3) Out-of-context techniques are taught with little if any consideration as to the legal ramifications on use of force (risk freedom and increased liability).

4) Little if any actual training is given on post-engagement measures with regard to first-aid and how to best deal with first responders (assumes things end happily in their favor).

Upon examination, it becomes apparent that improved performance in stressful situations requires sufficient attention be given to becoming well-prepared *and then skilled* in the pre-engagement, engagement, and post-engagement phases of an event.

PRE-ENGAGEMENT

"Let your enemies be disarmed by the gentleness of your manner, but at the same time let them feel, the steadiness of your resentment." – *Lord Chesterfield*

As preparations are the actions we take to get ready to implement the strategic objective, pre-engagement activity aims at becoming more skillful at managing the dynamic interactive moments leading up to an engagement.

How well we manage the pre-engagement stage very often has a significant impact on the outcome of the engagement, and this is especially true when it comes to stressful, time-pressured events.

With regard to self-protection, the task of pre-engagement skill becomes even more context-specific by identifying and understanding the attacker's mentality, goals, and approaches, as well as knowing the pitfalls of common victim mentality.

A criminal who intends to rob you has an entirely different motive than a bully who wants to humiliate you, or a predator who intends to sexually assault you, and responding to them is not one-size-fits-all.

Recognizing this helps us to practice more effective safety measures like de-selection (not displaying valuables or appearing vulnerable), blending in (dressing and behaving in ways that do not attract attention), rehearsing verbal confrontation skills, and training escaping tactics.

When approached by a potential criminal attacker, they seek to gain *proximity* (get close enough to pre-engage you), then simultaneously make their *approach* (conversation designed to occupy your thoughts) and get in *position* (close enough to attack) so they can physically attack (engage you).

Attackers come to the event mentally and physically prepared, and have a substantial advantage over someone who was caught completely off-guard with no pre-engagement protocols to follow. The average victim typically falls prey to the attacker's proximity, approach, and positioning set up, and find themselves face-to-face with a threat. At the very moment they ask themselves, "What should I say?" or "What should I do?" it is all over.

Most potential attackers will decide not to follow through if they are unable to establish the controls with which they have learned to successfully operate. When things aren't going their way, they tend to break off and select someone else.

Therefore, before I train anyone in fighting tactics, I train them in pre-engagement tactics (not advice, but actions): how to avoid/reduce people from gaining proximity, exactly how to maintain independent thought and self-control if approached, how to make it difficult for an attacker to gain the positioning they seek, and finally, how to recognize cues of impending attack that trigger last-resort options to escape.

If unsuccessful at dissuading the attacker (and this happens when an attacker has decided to attack no matter how a victim responds) then at least the intended victim receives that attack on the best possible mental and physical platform on which to physically resist. Pre-engagement skills can have a powerfully decisive effect on the outcome of a violent interaction.

In contrast to criminals who seek to fatten their wallet by emptying yours, bullies seek to boost their ego by humiliating victims. The typical eye-contact challenger asking, "What are you looking at?" creates a false offense and escalates it into a challenge to fight: "What? You want to fight with me?"

I train people how to disengage from this, but when circumstances force them to manage the bully, I teach them to use the recognition ploy. "Sorry, but aren't you a friend of my brother, Bob?" It gives a plausible and innocent reason to explain why you were looking at him (now he can't say you're calling him a liar) and since everyone likely knows someone named Bob, just asking the question causes him to consider if he might be picking a fight with the brother of one of his friends. "You look so familiar. I wasn't trying to call you out or anything - besides, I don't want to fight, I kinda suck at it and you'd beat me up!" This appeases his ego and allows him to save face – plus he'd look more like a jerk for attacking someone who already backed down.

If the recognition ploy doesn't work and the bully persists with another challenge, I ask them, "Can we talk?" (which implies I want the event to remain verbal and to go on), and in the moment their mind considers my question, I initiate an attack and strike them first, setting up my escape. The training continues, dealing with "what if" contingencies, but you get the idea. Without rehearsing pre-engagement tactics, bullies will almost always have the upper hand over their victims.

What motivates the attacker, and what are their modes of operation?

- Criminal (predatory) attackers are principally motivated to take your money, possessions, resources, or by sexual assault motives.

- Personal (emotional) attackers are principally motivated by anger, fear, despair, desperation, ego, pride, reputation, greed, hate, insecurity, jealousy, or revenge.

- Instrumental (compelled) attackers are principally motivated by acceptance, initiation, defense of self/others, duty, ideology, politics, religion, the urge to test oneself, territorial, or status related issues.

- Anti-social (frustration) attackers commit acts of violence by opportunity (random selection of victims does not imply without motive); this includes attacks committed by drunk/drugged people.

- Mental illness (psychotic) attackers experiencing a lapse in their connection to reality could be paranoid, delusional, schizophrenic, or be suffering from other functional defects.

While these examples illustrate my approach to pre-engaging with attackers, it is useful to recognize the principles at work and to apply them in everyday confrontations. Pre-engagement is something we do in everyday life, whether we are asking for a date, going to a job interview, making a presentation, avoiding a criminal attack, or any number of real-life scenarios in which we wish to perform well.

ENGAGEMENT

"Invincibility sure is elusive." – *Tom Sotis*

Over the course of history, thousands of books have been written on engaging in various forms of combat, because technically speaking, engagement is a topic that could fill several sets of encyclopedias. This book intends to present the elements of knife combat as a guide for people to achieve invincible performance with whatever they are attempting to do in life.

Let us first ask the question: Is it possible for someone to become invincible? And, of course, most people would agree that invincibility is a figurative concept, something of which humans are not literally capable. And they would be wrong.

"Invincibility, the state of performing physical feats with flawless execution, is indeed rare but it is perfectly natural and achievable." – *Tom Sotis*

I state, unequivocally, that it *is* possible to be invincible. I do not believe in the figurative concept of *sustained* invincibility (where one person could defeat all others) but I am certain that the literal reality of *momentary invincibility* exists and is typically achieved by high-performance individuals under high-pressure circumstances.

High levels of danger/risk seem to activate more frequent incidences of invincibility, and I categorize them by context:

- "Competitive invincibility" occurs in high-level sports events by top athletes.
- "Combative invincibility" occurs in mortal combat by trained combatants.
- "Survival invincibility" occurs in life-threatening circumstances by everyday people.

The purpose of Invincible Coaching is to increase one's ability to actuate on command, and to sustain an invincible state of being.

When engaging with an enemy, the three stages of any action: initiation, execution, and recovery can occur in the fraction of a second. Successfully countering such actions permits virtually no margin of error in one's mental control and physical readiness.

Elements of Invincibility:

Motivation – know your why
Mindset – commit to your strategy
Vision – adjust your window to the world
Attention – the recognition and filtering of movement cues
Intention – the *one* thought you decide to hold in your head
Posture – the combined quality of structure and balance
Breathing – tension in abdomen, 3/5th reserve with circular breathing
Action – explosively triggered by their action
Recovery – re-stabilizing one's mind and control over centerline

In order to actuate a state of invincibility, we must bring all of these elements into play, so I have developed a simple way to practice it.

Actuation Exercise: Fix your eyes and your posture as you form an intention, using tension in your abdomen to maintain it until triggered to release it. *In case you didn't recognize it, this is the key to the proverbial kingdom!*

Causes of Vulnerability:

Motivation – your values are in conflict
Mindset – doubt in yourself or strategy
Vision – during blinking or saccades
Attention – external distractions
Intention – internal dialogue
Posture – during changes in posture
Breath – At the end of a breath
Action – When chambering before striking
Recovery – intention guarding centerline

In the last chapter, we covered the elements of Motive. In this chapter, we explored Mindset, and established the remaining elements of State of Invincibility, which are covered in the next three chapters.

Attack and Defense are not like two sides of a coin where the fighters take turns doing moves on each other like flipping a coin heads-over-tails. Fighting is actually more like flipping a coin that has heads/tails on both sides.

On one side, you have Attacking (*initiating*), which simultaneously contains and combines offensive/defensive expectations and actions. To attack with no consideration for one's own vulnerabilities is to invite disaster, hoping to best the moment by chance instead of skill.

On the other side, you have Countering (*responding*), which simultaneously contains and combines defensive/offensive actions. To defend without employing some means of halting further attacks is to lose by attrition.

Fighting is an oppositional event where both fighters simultaneously attempt to impose their moves upon each other, i.e., the flipping of the proverbial coin.

Generally speaking, successful Attackers don't target hard-to-get targets with ineffective weapons; rather, they attack available targets with the most effective weapon.

Because an attacker "counters the opportunities provided" it can be posited that the "*initiating* movement" was actually a deceptive movement by the "*responder*" who invited it, thus dissolving the concept of attacker and defender into a single symbiotic entity that expands (pressure decreases) and contracts (pressure increases).

This single, symbiotic entity cannot adequately be perceived within the context of a series of attacks and counters, but more as a continuity of one mutual-counter-action unfolding into the next mutual-counter-action. At some moments combatants are in contact and at other moments, they are not. The cycles of continuity continue until one or both make an error and thus, a winner and a loser emerge.

POST-ENGAGEMENT

"It ain't over till it's over." – *Yogi Berra*

If we are to achieve the best possible outcomes from violent encounters or events, post-engagement measures are a vital part of strategy and preparation. When actively involved in violent fighting, auto accidents, active shooter, or other human-crisis events, there tends to be one of four immediate outcomes: you end up either dead, injured, arrested, or back home.

Dead, in the Morgue

It's a pretty safe bet you aren't planning to be killed, but you can plan for your unexpected death. Of course, if there are cameras, witnesses, DNA, or other form of evidence, those might help police to identify your killer, but you have little control, if any, over those factors. You do have control over three main strategies, and they are 1) put all your personal affairs in order, 2) put your business affairs in order, and 3) get life insurance. Knowing your loved ones are financially secure and protected if something happens to you is a weight off your mind, and the alternative is a distraction that you don't want to carry into a moment of crisis.

Being ready to die isn't at all the same as wanting to die—it means you can do it well.

Injured, in the Hospital

Because the probability of injury is high for participants and sometimes bystanders, it is a good idea to have knowledge of basic first-aid skills in case it becomes necessary for you to save or treat yourself, your loved ones, bystanders, and sometimes even attackers. If you are allergic to medications, it is a good idea to have that info on you for EMT's to discover (i.e., penicillin listed on an ID bracelet).

As well, it is advisable to be cautious about statements you make to Emergency Medical Technicians on the scene, and to doctors and nurses in the hospital. Your right to privacy and confidentiality are very strong, but it is a mistake to assume that they are airtight. And finally, ideally you have good medical insurance.

Arrested, in the Jail

I am about to make some generalized statements that are typically subject to many deciding factors. If you attack someone and injure them, there is a likelihood you will be arrested. If you injure someone in a mutually-agreed-upon fight, there is a good chance you will be arrested. If you are attacked and successfully defend yourself, there is still a possibility that you will be arrested. If you are attacked and are injured by your assailant, unless you broke some other law, it is unlikely you will be arrested, but still possible (because you were present at a bar brawl or protest, etc.)

Home, upon Arrival

After a violent event, our memory can be affected in a number of ways and because important details can be lost just hours afterwards – and even more so by the next day after having slept – it can be advisable to write down or videotape as many details as you can while they are still fresh in your mind. It can be helpful to start by answering the standard investigative questions pertaining to the event: who, what, when, where, why, how, and how much? It is usually a good idea to obtain a copy of the police report, and to immediately seek legal representation if anything negative arises.

Home, but Not Over

Getting home beats ending up at the morgue, hospital, or the jail, but it doesn't mean the course of events has concluded. You could be in danger from your attacker's family/friends or a vengeful attacker. (Side note: this is why I always advise my clients that whenever they can, they should get or take some form of identification from anyone with whom they've had a violent confrontation. It is preferable that you know who they are and where they live, but that they do not know who *you* are or where *you* live). The other forms of collateral damage are not to be dismissed either, as there is potential for consequences such as a civil lawsuit against you, for your business to suffer a public relations problem, for personal problems to be brought into your relationships, and even for mental health issues to arise (sleep, substance abuse, anxiety, etc.).

> "We do not learn from experience ... we learn from reflecting on experience." – *John Dewey*

Post-Engagement Reflection

Reflection is a natural part of processing our experiences and if you are experiencing psychological struggles, therapists are experts in asking the right questions to assist you in better managing the aftermath of violent events. Apart from psychological issues and more from the standpoint of practical learning experience and gaining some tactical wisdom, here is a small sample of the kinds of questions to reflect upon:

- What sort of factors led to the event?
- How might they be mitigated in the future?

- What did I learn in the pre-engagement stage?
- What did I learn in the engagement stage?
- What did I learn in the post-engagement stage?
- What lessons did I learn overall?
- What steps should I take now?

" Honest self-reflection opens your mind to reprogramming, change, success and freedom." — *Unknown*

INNER ARENA

"We must embrace pain and burn it as fuel for our journey." – *Kenji Miyazawa*

For people of substance who hold inner visions of courage and achievement, AMOK! offers both the opportunity and a safe arena in which to challenge the spirit. When you fight, you learn about yourself because fighting tests your beliefs and values not only about yourself, but also about your wider worldview.

"A real fight doesn't begin until the moment you want to stop but you must continue." - *Tom Sotis*

Before that moment when you want to stop, you can call adversarial contact with someone else a number of different things: a match, a bout, a tune-up, an ass-kicking ... but if you are not forced at some point to continue when what you really want to do is quit, then it never evolved into a true "fight" for you.

When fighters talk about "digging down" or "digging deep" to find the strength of will to go on, they are speaking about the reserves they have stored up from training. Every time you fight in training, you add a little bit more to your reserves; every time you give in, you lose some.

When I speak of fighting, I mean not quitting when you feel:

- incapable of meeting a task or goal
- unworthy for failing a task or goal

- bored from necessary repetitive tasks
- hungry from cutting weight
- unhappy with your performance
- guilty for lack of self-discipline
- restless and/or anxious before training
- lonely during solo workouts
- social isolation due to training regimens
- doubting your motivation/commitment
- afraid to disappoint others
- disappointment from not winning
- disillusionment of winning
- betrayal by coaches and people you trusted
- the accumulated physical exhaustion and pain reducing you to fetal position

With each of those challenges to our inner selves comes equal opportunity to feel capable, worthy, curious, satisfied, happy, contented, confident, resilient, independent, safer, and perhaps a little wiser.

This is the "polishing of one's spirit" or seeking perfection of character by embracing and overcoming emotional and psychological weaknesses. Warriors live this way not because it is easy, but because it is powerful and it provides a means of personal achievement.

CHALLENGERS

> "Challengers are like mirrors; they force you to see yourself." – *Tom Sotis*

The two different types of people I have fought are challengers and attackers. Where you agree to fight challengers, attackers give you no choice. I classify challengers into three groups:

amateur challengers seeking status, professional challengers seeking experience, and private challengers seeking combat.

Amateur challengers do so in an effort to boost their public reputation. These situations typically occur at seminars where the challenger waits to find something in particular they believe they can counter, and then asks you to prove it using them. Because their life is at stake, I always encourage people to challenge the *information* because it is much different than challenging the person. Politely asking, "Can you show me how this works in real time when we go full-on?" is far different than rudely saying "You couldn't do that to me." If the challenger wins, he has everything to gain and nothing to lose (except pain) but since he is the underdog, there is little risk on his part. For the instructor, everything is on the line: his health, his professional reputation, and perhaps his only form of income to support his family.

Status challengers typically opt for the relative safety of an environment with witnesses. That doesn't mean there isn't a definite danger, but the loser won't end up in the gutter with his throat slit. It is a mistake to underestimate anyone, and that goes for even amateur challengers. Just because someone may be unknown outside of his/her neighborhood doesn't mean they aren't a dangerous opponent. Having taught over 1000 public seminars, I've had more than a few status challengers; as of this writing, I have not yet been bested.

Professional challengers are instructors of other methods who have become friendly and want to duel more for the purposes of experience and research than to hurt one-another, so they agree to go full-on until first blood, or until someone is hurt. Then, they take care of each other; they talk about what they learned. And out of mutual respect, never talk about it again.

Private challengers are extremely rare and very dangerous, as they aim to injure, disable, or even kill. All over the world there are sharks in the water, and if you swim in their territory, they will indeed come for your blood. The most dangerous challengers will approach you privately and politely request that you agree to engage with them somewhere private and appropriate for such an event. If you choose not to take advantage of that courtesy, from that moment forward, the situation is considered to be an active challenge whereby the challenger may attack at any opportune moment, and very likely, at a time when you are less prepared.

They do this not because they are observing some "knifer's code" (though I suppose anything is possible), but because for them, it is about testing their skill, so they want to fight with you at your best. If they just wanted to kill you, they would ambush you. Anyone who participates in such an event is unlikely to admit it, so it is both pointless and impolite to ask that of someone's history.

ATTACKERS

> "The same hammer can drive different kinds of nails." –
> *Tom Sotis*

Attackers differ from challengers because violent attacks are not mutually-agreed upon events. Attackers employ every cunning tactic they can to gain an unfair advantage over an intended victim. Having an awareness of *why* people attack others assists in recognizing those situations as they develop so that we have a better chance of avoiding it, escaping it, or managing a better outcome.

When you know what motivates an attacker, you have a better chance of communicating, behaving, or taking action in ways that can be to your advantage. I classify attackers into these categories: personal, ego-based, anti-social, compelled, criminal, predatory, and terrorist. AMOK! offers a hammer for each one of these "nails".

Personal attacks are emotionally-driven. Anger and hatred are very powerful motivators, and arguably, the most common. People have disagreements, and sometimes verbal arguments escalate into personal violence. As well, people suffering from despair or fear, or who are torn by desperation can feel compelled to attack.

Ego-based attacks are insecurity-driven, and are personified by the stereotypical bully seeking to boost his ego and hide his insecurity at the expense of someone's humiliation. Other ego-based attacks are spawned by an insult to someone's pride, a perceived threat to reputation, jealousy, and envy.

Anti-social attacks are frustration-driven. When inner pressures build up and, in many cases, behavior becomes less inhibited by alcohol and/or drugs, people more easily take offense and respond aggressively. Some people will intentionally provoke someone to attack them, while others will look for someone to attack. Remember, all attacks are motivated, even when the victims are randomly selected and the attack seems not to have served any purpose beyond face value. "Senseless" does not mean without motive.

Compelled attacks are belief-driven. Some people feel compelled to attack others over differences in their beliefs about religion, ideology, justice, moral obligation, duty, or revenge.

Criminal attacks are objective-based. Some attacks are motivated by valuables (money, possessions, resources), other times criminals are motivated mainly by the physical assault, and sometimes, by both. Attacking others is a common form of gaining group acceptance, as rites of initiation among gangs, and as a matter of protecting territory and status.

Predatory attacks are sexually-based. These attacks are motivated by unrestrained sexual urges and are universally regarded to be among the most heinous and despicable of all crimes.

Terrorist attacks are politically-based and are attacks made by those who use unlawful violence and intimidation, especially against civilians, in the pursuit of political aims.

READINESS

"You are only as good as you are right now, in this moment."
– *Guro Dave Gould*

We have covered the spectrum of elements required to generate the state of temporary invincibility and, as remarkable as it is elusive, unfortunately, invincibility is not sustainable.

In between those blissful moments, we exhibit varying degrees of readiness (capability) which is normally in a state of flux as an encounter unfolds. While readiness is not constant, we train to raise the bandwidth of our readiness to function at its highest level of sustainable performance, thus increasing the likelihood and frequency of achieving invincibility.

We raise our bandwidth of readiness by developing the synergistic qualities of physical control, awareness, and vision. More details will be discussed in the chapters focusing on tactics and actions.

We strive to optimize physical control:

- while balanced, off-balance, and during balance recovery
- during controlled, labored, and uncontrolled breathing
- through good posture, efficient movement, and use of leverage

We strive to optimize mental control when:

- acting on early recognition cues with intentional decisional actions
- reacting to late recognition cues with unintentional reflexive actions
- hesitating at partial recognition cues with natural defensive actions

We strive to optimize our vision by:

• centering our adversary(s) in our field of view and tilting this field of vision downward to bring his feet and our flanks into view.

• not focusing our eyes on *any* object; focusing the eyes is a result of your intention to isolate a single object upon which you wish to concentrate, leaving you less able to detect cues of motion.

• not losing sight-picture by controlling blinking and saccades.

ON FEAR

"Ultimately we know deeply that the other side of every fear is a freedom." – *Marilyn Ferguson*

Fear is not real, danger is real; and while fear is *a* response to danger, it doesn't have to be *the* response.

We have covered how threats can affect our state of awareness, and now we will examine those thoughts as part of the three-way connection between our minds, our emotions, and our bodies.

Dr. Robert Sapolsky explains this connection in a brilliant TED Talks video – what I present here as the "Triangle of Fear". There are three main functioning apparatus, each affecting the other:

• *Reptilian* bodily functions: heart rate, respiration, temperature, urination, defecation, adrenaline, hunger, sexual desire, etc.

• *Limbic* emotional feelings: anger, fear, joy, sadness, surprise, contempt, sadness, and disgust

• *Neo-cortex* conscious thinking: thoughts, questions, assessments, problem-solving, and creativity

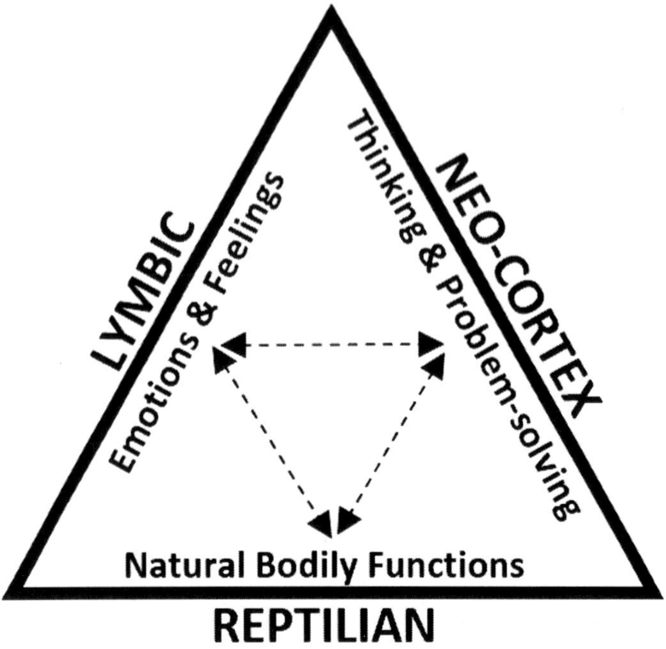

– Reptilian affects Limbic: Your body function makes you feel something (i.e., you become hungry and begin to feel angry)

– Reptilian affects Neo-cortex: Your body function makes you think something (i.e., you become hungry so you think about what to eat)

– Limbic affects Reptilian: Feel something that makes your body react (i.e., fear makes your heart rate and respiration increase)

– Limbic affects Neo-cortex: Feel something that makes you think (i.e., feel fear: think fight or flight)

– Neo-cortex affects Limbic: Think of something that makes you feel an emotion (i.e., think about bully: feel angry)

– Neo-cortex affects Reptilian: Think of something that makes your body react (i.e., think about bully: heart rate and respiration increase)

With all three apparatuses affecting each other, if you fail to control them, stressful situations can gain momentum and have the potential to become overwhelming. Stress inoculation is a process by which prior success under stressful conditions acclimatizes you to similar situations, thereby promoting future success.

When training retail store detectives how to arrest thieves (many of whom were intimidating and for good reason) I had to confront threatening individuals and groups on a daily basis. Soon, the apprehensions and nervousness that I experienced regularly at first began to subside, and when they occasionally resurfaced, I discovered, I recognized the thoughts and feelings to be the tentacles of my instincts.

This kind of experiential learning underscores the importance of realistic scenario training in developing self-protection skills, and is evidenced by the fact that virtually every police and law enforcement training program employs them.

We encounter a paradox when facing adversaries, especially intimidating ones, and it is: *I don't want to underestimate my adversary and I still want to feel confident that I can defeat them.* To resolve this paradox, I assume that my adversary is a better fighter than me; however, I am confident that I have the capability to capitalize on his mistakes well enough to win. And my adversary will make the first mistake the very first instant they move because I will employ deception, and they will underestimate me.

In the chapter on Engagement, I presented an exercise to actuate a state of invincibility: Fix your eyes and your posture as you form an intention using *tension in your abdomen* (tighten your lower stomach muscles) to maintain it until triggered to release it.

Sometimes, the reptilian responses of heart rate, respiration, and adrenaline can ramp up so quickly that they threaten to spiral out of control. You may feel your heart race, you may even begin "gulping air" (trying to breathe, but there seems to be no more room in your lungs to inhale), and your knees might shake.

Over time, I developed a default technique to quell any rising bodily functions by simply growling – not a voice-box growl, but a deep, low guttural growl that emerges from the pit of my stomach. This type of growl (tension in your abdomen) seems to collect the released adrenaline and stop it from rushing through your body (thus the saying about *feeling a fire in one's belly*). In addition, your knees won't shake, and it serves to stabilize your breathing because having exhaled, your body now wants to inhale again.

To improve your performance when faced with danger, *growl away your fear.*

TACTICS

DECEPTION

"All warfare is based on deception." – *Sun Tzu*

Deception can be broadly defined as the manipulation of appearances such that they convey a false reality. They are the means used to exploit your opponent's thinking, causing that person to misjudge or miscalculate what they see before them.

> "To avoid falling for deceptions, remember: If his lips are moving, he's lying; every action is a distraction." – *Tom Sotis*

The mind can only focus on one thing at a time - make your opponent think what you want them to think by using questions to funnel/focus their mind. Make them look where you want them to look, and let them see only what you want them to see. Therefore, use words to distract the mind, and your body to distract the eyes.

Deceptions include, but are not limited to: readiness, unpredictability, looking, telegraphing, false-read, psychology, distractions, and probing.

Readiness

> "Present your weaknesses as strengths and your strengths as weaknesses." – *Tactical Wisdom*

Readiness is when you present no apparent defensive capability, behaving in such a way that it appears you have no awareness or ability to defend yourself, when in fact, your counter-action is poised like a hidden archer, waiting to unleash an arrow.

Disguising readiness is the key to gaining an advantage through the element of surprise. It deceives your attacker into underestimating your state of readiness, and is designed to lure them into executing a hasty attack as they hope to capitalize on the opportunity before it evaporates. For example: wearing earbuds with head bobbing to music while looking out the window, but earbuds are actually turned off and you are observing via the reflection in the window.

Another ploy is feigning an injury or disability before or during an altercation. The disability might be sickness, injury, physical impairment, mental impairment, deafness, blindness, being crippled, distracted, sleeping, drunk/high, or any other creative measure that conveys a false reality.

Unpredictability

> "Tactically-speaking, predictability is a leading cause of death." – *Tom Sotis*

Deceptions are actions you take to exploit your opponent's thinking, whereas being unpredictable involves measures that you take to make it harder for an opponent to know what to expect from you.

A combative rule of thumb for unpredictability is to avoid doing anything in threes. Change the geometry of your guards and don't repeat any guard three times before changing. You can execute an attacking tactic a second time in a row, but not a third time. Once is an instance, twice is a happenstance, and thrice is dead.

Conversely, you can use predictability to set traps by purposely repeating an action, psychologically inviting your adversary to capitalize on it. When they think they are seizing a momentary opportunity, they are actually stepping into your trap instead.

Looking

> "The key is to look like you're looking at something without looking like you're looking at it." – *Tom Sotis*

The importance of controlling our vision was mentioned in the chapter on Engagement, and the means of visual control is covered in the upcoming chapter on Vision.

In a different way than the mechanical controls employed for optimal visual acuity, we can use our eyes tactically to deceive our adversary. An astute adversary will notice even the most minute fluctuation of your eyes. When you are reading your opponent and believe you have detected a weakness, typically something drew your attention, and your eyes, no matter how minute or fleeting, move toward that object. This is exactly the kind of "tell" poker players seek to hide themselves, but observe in other players.

"Show me your eyes and you may as well show me your cards." – *Doyle Brunson*

Looking is purposefully allowing your opponent to see you glimpse a particular target in order to deceive them into believing they have read your attention. This sets them up nicely for fakes when you seemingly attempt to go for the target and they counter, but you are already circumventing their action.

The deceptive looking exploited his thinking, while the fake attack exploited his reflexes. Deception and faking can be executed independently, or together.

Telegraphing

> "The rule is this: you spot a man's tell, you don't say a f*cking word." – *Mike McDermott (Rounders)*

Similar to looking as a means of deception, telegraphing seeks to deceive through a barely imperceptible shifting of your weight, minuscule adjustment of your frame, a slight angling of your blade, the repositioning of your empty hand, the angling of your foot, and other such small movements that surreptitiously reveal your plan of attack and convince your adversary that they have read your intention. Then, you switch to attacking another target. Telegraphing and Looking are commonly combined to convince an opponent of a false intention with subtle cues. Telegraphing also combines to illicit False-Reads when seeking a desired response to circumvent.

False-Read

> "The best way to crush someone's spirit is to first raise their hopes so they fall even harder." – *Tom Sotis*

To false-read is to strike in such a way that if your adversary successfully blocks (which conveys to your opponent that they have succeeded) upon impact, you deflect, rebound, or push-through to strike what was your intentional target all along.

You might also reverse the application of a false-read when you are struck and you fake being injured more than you actually are so as to deceive your opponent as to your state of readiness.

Psychology

> "One need not destroy one's enemy. One need only destroy his willingness to engage." – *Sun Tzu*

Your adversaries will make use of propaganda, threats, and other non-physical techniques during confrontations or fighting in order to mislead, intimidate, demoralize, distract, or otherwise influence your thinking or behavior. I have characterized some deceptive behaviors so as to assist in recognizing them, thus reducing or completely nullifying their effect, in which case, your adversary's failure to affect you will produce a negative effect on them. I classify these adversaries as Manipulators, Berserkers, Weirdos, Psychos, Latecomers, Predators, and Escalators.

Manipulators aim to put thoughts into your mind to affect your emotions. They may threaten to reveal secrets, or invent lies to cause embarrassment. They might lie to cause jealousy, confusion, or cause you to distrust another. People will make you happy to gain your trust before betraying you, or give you a gift before stealing from you. Others may attempt to make you angry so that you don't think clearly, and may become more susceptible to manipulation.

Going *berserk* is another deception where someone *seemingly* loses all control and goes on a destructive rampage designed to give the illusion that they are not thinking clearly. Berserkers use chaos as a ruse to buy time, change the dynamic, or disguise their intention.

Strange, *weird*, and unexpected behaviors can catch people off-guard and destabilize their mindset. A good example of this can be seen on YouTube where a guy picks a fight with a big tough guy and when the guy puts his hands up ready to fight, the challenger rips off his tear-away pants, revealing a tiny thong and asks, "You want some of this?" Every guy who was ready to fight ran away.

Psychos go to extremes to convince you they are unmanageable. Ordinarily, mentally healthy people don't purposely hurt themselves, so another common form of deception/manipulation is when someone inflicts pain on themselves to demonstrate their lack of fear of pain, their tolerance to it, and to convince their opponent that they are mentally-disturbed.

Latecomers employ another form of deceptive manipulation. They agree to the place and time for an engagement, after which time (with the stipulation permitting up to one hour to be late) if either one failed to show, they automatically forfeit the engagement. Then, they purposely show up moments after the one hour window, when the match has been forfeited and your opponent's state of readiness has been expended. In this case, apologize for keeping your opponent waiting and tell him you are ready to engage. He will be angry that you kept him waiting and his state of readiness will be very difficult to re-establish after having released much of his mental energy when he thought he had won by forfeit. When two of Japan's greatest swordsmen agreed to duel, Musashi used this technique to defeat Kojiro.

Predators employ another psychological lever by making direct threats of physical pain, disfigurement, disability, and/or sexual assault. These threats are then amplified by making you watch while they do the same to your loved ones. Since deception is designed to *elicit a response in you and from you,* learn to show no emotional response. Ask them, "Does that usually work for you?" This helps you to control your emotions, doesn't give them a measurable response, and implies that your opponent's tactic has failed.

Escalation is a deception/manipulation intended to raise the stakes such that the challenger decides it is not worth it. If someone approaches you in anger and intends to hurt you, offer them the wind of hate and declare your intention to take their life. If someone approaches you with the wind of hate intending to kill you, offer them the spirit of cruelty and declare your intention to leave them crippled and disfigured in lifelong agony and suffering.

More than once when confronted by a thug, I looked them right in the eye and in a measured tone I made it clear: *I do not want to fight but if you touch me, I'll kill you – one way or another.* Unless the adversary is willing to instigate a life-or-death encounter, they will let you off. This can be an effective manipulation *only* when you can effectively back it up and are ready to escalate things to mortal combat. Never bluff using this tactic.

Distraction

> "I've trained all my life not to be distracted by distractions."
> – *Nik Wallenda*

Some distractions come from the environment, or from people in the environment: a car alarm goes off, parking lot lights come on, someone runs past you, a bottle is thrown. It could be a big distraction, or multiple smaller ones. Any one of a hundred things can cause a momentary distraction on which either of you might capitalize if the other gets distracted.

Tactical distractions are actions taken specifically to redirect your adversary's momentary attention, both visually and mentally. Distractions have to be reasonable—not out of context, clumsy, or convoluted—and they must be well-executed.

Looking is the action used to divert their gaze. When you have locked eyes with someone and then purposely look at something else, their eyes tend to follow to see what you are looking at. Asking a question triggers their mind to answer it and serves as a momentary mental distraction. When faced with circumstances where I must strike first in order to escape, I will usually ask, "Can we talk?" and initiate my strike as the last syllable passes my lips. The question implies that I want the encounter to remain verbal and continue – which is the opposite of what is about to occur.

There are many more distraction techniques limited only by one's creativity in any given circumstance, but it is important to employ distractions properly, and not to resort to trickery as a substitute for confident command over your adversary.

Here's a distraction that we have used to more easily capture criminals on the street. When the target walked towards us, we would walk toward him, but one of us would walk about ten feet ahead of the other so it would appear that we were not together.

As the man in front passed by the target, he would rudely bump into him so hard that the target would turn around, angry to confront our man. The front man would stop, give the target a dirty look, and ask "WHAT?" At that moment, the target's attention is quite concentrated and focused straight ahead and never sees the second man walk up right behind him ... Game over.

Probing

> "Deceptions are designed to elicit a response in you and from you." – *Tom Sotis*

Probing is a vital tactic to develop because when done correctly, it gives you a glimpse at your adversary's reflexes, which are an involuntary and nearly instantaneous movement in response to a stimulus.

Under certain conditions, quickly threaten your adversary with a false attack and observe their ingrained physical response. Did they overreact, underreact, delay-react, or fail to act? If you replicate the same conditions and offer the same stimulus, there is a high likelihood that you will trigger the same response which then becomes easy to circumvent or counter.

As a combative rule of thumb, is unwise to probe and then immediately attempt to capitalize on a successful probe because your opponent may be aware you read into him, and ready himself for another attempt which he now knows will begin with a feint. Therefore, after making a successful probe, immediately change your stance or frame, or act in such a way that forces his mind to remain attentive to the present moment, thus failing to process that you have successfully probed him. Now that your foe is occupied with the present moment, you can set about recreating the conditions you intend to exploit.

When someone has successfully probed me and I have inadvertently given my adversary a glimpse of my reflexes, I alter my stance, posture, or guards sufficiently so as not to offer my adversary the same conditions. I might also pretend that I didn't notice, and set a trap myself so that when they attempt to capitalize on my reflexes, instead of being a moment behind in reacting, I'll be ahead of the moment – reading his cues, and acting on it.

MOVEMENT

"Float like a butterfly, sting like a bee." – *Mohammad Ali*

Tactical Movements are actions that are the primary means to an end, where the movement itself is designed to produce a tactical advantage (the footwork used to achieve it is technical movement).

Tactical movement requires an acute awareness of the distance between your reach, and your adversary's reach (cusp).

In unarmed combat, because a human's height equals the reach from fingertip to fingertip, with a few simple exercises, one can learn to better judge someone's reach by their height. In unarmed combat, two people of the same height have equal reach.

Armed combat requires an even greater acuity in judging reach because the height factor of unarmed combatant combines with potentially varied lengths of the involved weapons to establish height (true cusp).

A combative rule-of-thumb in fighting is to stay just far enough outside the cusp so they can't reach you, but not more. This forces an opponent to take a step in order to close the distance while minimizing the distance you must close to reach them. When I first trained with David Gould, it was like fighting a ghost. He was right in front of me, I'd strike as fast as I could and it would appear that he had hardly moved if at all, except like a bad dream, I'd miss completely, and before I could recover my strike, the tip of his weapon would exact payment in pain for my mistake.

Initially, the footwork of novices is quite pronounced, using a lot of movement when moving outside the reach of an adversary's attack and then getting back in close enough to reach them. What you want to notice is the distance they moved their center of gravity during the counter. To ensure they don't get hit, novices tend to move too far back from the cusp which lengthens the distance/time it takes to deliver their counter-strike. Experienced fighters with a more acute ability to judge the cusp use less footwork with subtle shifting to stay *just a hair* outside the reach of their opponent, and effortlessly regain that distance/time while their adversary's strike is still in recovery.

GUARDING

> "Protect yourself at all times." – *First rule of boxing*

"Keep your guard up" is good advice, but so is "open your guard up" because in fighting, context is everything. What is true/good in one circumstance is often untrue/not-good in another.

Now I will reveal a concept so powerful that nobody will ever attack you again. Ready? *Nobody simply attacks you, they counter the opportunities that you provide*, whether or not you are conscious of, and purposeful with them. This seemingly obvious statement has profound implications.

Therefore, it becomes essential to use your guards very purposefully towards controlling, or at least limiting, how your opponent attacks. In this way, your deception and purposeful guards draw them to take an action which they believe initiates an attack, when really, they are acting on (countering) the opportunity your offered.

Moving guards invite direct attacks when your guards are presented and in motion. *Stationary guards* invite either indirect or direct attacks, depending on the placement of your guard.

CONFIGURATIONS

> "In fighting, your preferences are a product of, and a reflection of, your weaknesses." – *Tom Sotis*

People often ask me if it is better to hold your knife in front, or to the rear. Or they ask if it is better to grasp it in forward or reverse hold. The answer to these polarized questions is *yes*. Anyone who advocates for one position over the other can only ever be half-right because *tactical context dictates* which are the most advantageous positions to pursue in the moment.

To maintain a position of tactical disadvantage out of personal preference or some kind of *combative bias* is perhaps understandable from a psychological standpoint of operating in one's comfort zone, but it seems both an unnecessary and unacceptable limitation when your survival is at stake.

A good example is right-handed people who have trained in kickboxing who operate with their right hand held to the rear. On the first day of knife fighting, they hold the knife in their right hand to the rear because that is what they know and do best, but lacking equal reach to a forward-hold opponent and getting picked apart, very quickly the smart ones begin experimenting with changing their stance. No matter how much you point it out or explain why, and even when they agree, for some reason some people will prefer the rear stance right into their grave. And from them, I originated the saying:

> "You can lead a horse's ass to knowledge but you can't make them think." – *Tom Sotis*

Personal combat is an ever-changing dynamic whereby you can rarely dictate the terms by which it unfolds. We learn to operate with *combative rules of thumb* which are overruled and dictated by the demands of the moment. Some moments permit you to select your ideal choices, but many do not, and force you to effectively adapt or die.

Lead Foot

When there is some distance between us, typically I choose to hold my knife in my right hand, with my right foot forward. This offers maximum reach to my adversary, my knife creates a shield of protection they must penetrate, and my empty hand is positioned behind the knife, acting like a safety net.

Conversely, when we are in close contact, I normally choose to have my right foot to the rear. This makes it harder for my adversary to attack, suppress, or grab my weapon-hand while also making it easier for my empty hand to strike, block, parry, grab, trap, lock, and so forth.

Ultimately, it becomes important to be able to effectively operate from both positions and to effectively transition between them.

Weapon Hold

Some people advocate forward hold (blade comes out of the top of your hand between the thumb and index finger) while others advocate reverse hold (blade extends out of the bottom of your hand from your little finger). Arguing for the tactical superiority of one hold over another is tactically short-sighted.

Mechanically-speaking, we can slash and thrust 360° in either hold. In forward hold, some angles are easier to make than others, with the opposite ones being true in reverse. Either way, we need 360° coverage so neither hold has a tactical advantage.

If there lies any inherent advantage to either hold, you will find it in application. Forward hold is "more natural" because slashes are like empty hand slaps and thrusts are like finger-poking (natural everyday type of movements), whereas when striking in reverse hold, the slashes are like elbow strikes and thrusts are like wrist-hacks (neither are everyday type of movements).

Weapon Hand

Preferring to hold my weapon in my either hand would, as circumstances change, court potential disadvantages, so I prefer to hold my knife in whichever hand is best suited to employ it. I am right-handed, so naturally I tend to hold my weapon in my right hand, but I will transition my weapon to my left hand when it presents a tactical advantage over keeping it in my right (i.e., if my adversary managed to immobilize my arm in a clinching-type of struggle) or perhaps if my weapon hand/arm is injured.

Transitioning is not a skill to ignore in your development because moments of vulnerability may demand it, so they must be secure, smooth, and ideally, hidden from view or accompanied by distractions. Tossing your knife back and forth between your hands is for the movies; it is not a sound handling tactic when your life hangs in the balance.

Weapon Length

When there is sufficient distance between fighters, swordsmen enjoy a reach advantage over knifers because swordsmen can strike with comparatively little risk from the knife. When there is a very close distance between fighters, knifers enjoy a reach advantage over swordsmen because knifers can strike with comparatively little risk from the sword. So as not to fall prey to, and to be able to exploit to one's advantage, we must learn to fight up close with a sword, at a distance with a knife, and how to effectively transition between ranges.

For this very reason, I have a tactical appreciation for machetes. They operate at both long-reach and close-reach distances and are the "ideal ratio of length to weight for humans". The only evidence I have to support this opinion is that, in spite of its many variations, machetes are the universal worldwide tool of humans who cut sugar cane (about the same as bones) all day long; they must use a tool that is not too light, or they have to chop too many times. If a tool is too heavy, it becomes too tiring to chop with. If too long, each stroke takes more effort to recover from, and if too short, the range of motion required for an effective stroke is difficult to achieve. As form follows function, machetes are mankind's universal tool.

These are the major tactical aspects of configurations appropriate to this book. I have not included changes that are more technical in nature, and venture into details better suited to a technical manual.

ATTACKING

> "Always attack. Even in defense, attack. The attacking arm possesses the initiative and thus commands the action. To attack makes men brave; to defend makes them timorous." - *Steven Pressfield*

Of course, every fighter seeks to present no opening for their opponent to attack, while simultaneously seeking to recognize a cue of vulnerability in their adversary.

In reading your adversary, it is vital to filter all incoming information, giving attention only to the most relevant cues so as not to become distracted by irrelevant stimuli, all of which tax your mental resources.

In the chapter on engagement, we covered major causes of vulnerability which are revealed during eye movements, during changes in posture, at the end of a breath, when chambering before striking, and in failing to effectively guard centerline. In exploiting those vulnerabilities, attacking tactics are classified by one's *intention* to act (the desire/idea) which is then achieved by (a) physical action(s). Tactics are "issued individually" meaning, by a single mental command, each of which may hold many physical variations. The attacking tactics are executed by ambush, a single strike, a combination of strikes, by faking (feints), by grabbing, and by trapping.

Ambushing

The intention to ambush an adversary is to attack by surprise, attempting to offer the adversary no expectation or indication in order to "catch them with their guard down", thus creating more advantageous conditions for the attacker.

Tactical ambushes rely mainly on cover, camouflage, and stealth to achieve superior positioning; this tactic is typically used when an adversary expects an attack.

Deceptive ambushes rely mainly on use of a ruse to gain or exploit the adversary's trust, such that they have no expectation or means of detection, rendering that adversary far more vulnerable and less able to thwart the attack. Most ambushes employ, to a greater or lesser extent, a combination of both deceptive and tactical measures.

Single Attack

The intention to deliver a *single* decisive strike is, arguably, the most common form of attack. This occurs when a swordsman or knifer attempts a single strike, when a stick fighter takes a swing, when a boxer delivers a single punch, when a karateka makes a single kick, and so forth. Single strike attacks are delivered with a quick, jabbing action that focuses on the speed of retraction, or with hard powerful actions that focus more on the power of extension.

Single strike attacks can be delivered with moments in between (*Now* and *Now*) or in a series of repeating moments (*Now* and *Now* and *Now* and *Now*) which is not the same as a combination of blended actions (*NowNowNow*).

It is important to note the difference between delivering a series of individual strikes and a combination of strikes. It may seem, in many cases, to be a distinction without a difference. In some cases, practically-speaking, there is no discernible difference. On the other hand, in some cases, it does make a practical difference and your life will depend on being able to make that distinction.

Combination Attack

The intention to deliver a combination of two or more strikes is perhaps the second most common form of attack. This occurs when a swordsman or knifer attempts a two or three-strike attack, when a stick fighter takes two or more swings, or when a boxer delivers a combination of punches, etc. Combination strike attacks are delivered with a fluid action that focuses on the control, balance, continuity, and smooth transitions of one action to the next.

If we posit that there are eight directions of attack, as there are eight points around a compass, there are sixty-four combinations of two, providing the building blocks for all combinations of three or more. Of those sixty-four attacks, eight are most efficient (top eight), eight are least efficient (bottom eight), and forty-eight range in-between.

While the moment at hand will always dictate what is the most effective combination available, ideally, and all things being equal, combination attacks are delivered using top eight actions that force your adversary to employ the bottom eight actions in order to counter.

Faking Attacks

The success of a fake attack (feint) rests in convincing your adversary that you intend to do one thing, and thus elicit a response, while you intend and do something different. Everyone is familiar with "fake left and go right" because this tactic is so effective in getting past an opponent, or creating an opening where there wasn't one before.

Successful fakes are *pure of heart* (use true intention) and just as you trigger that action, you immediately halt, and engage in being *deceptive in mind* (change delivery) and *decisive in body* (effective mechanics). Although related and often combined, deception and faking are different. Deception is intended to exploit your adversary's thinking, while faking is designed to exploit their reflexes.

Grabbing Attacks

Another tactic is the intention to *grab* and/or suppress the arm/wrist in which your opponent holds their weapon (or the weapon itself) in order to restrict their ability to counter. In some cases, (especially empty hands combat) the grabs occur on clothing and parts of the body. This tactical configuration then compounds by employing other tactics, whereby once a grab is achieved, further actions are employed to capitalize on the momentary advantage. It is not uncommon in close quarters for both combatants to have grabbed each other, creating various configurations of stalemates resolved by the ensuing struggle.

Trapping Attacks

Finally, there is the intention to *trap* your adversary by forcing them to commit to a counter in which the next successive action on your part leaves them in an extremely difficult position to effect a successful escape, defense, or counter. Traps are the tactical version of painting your adversary into a corner. Trapping attacks typically employ a combination of deception, fakes, and grabs designed to first trick your mind, then your body, and then to ensnare you.

COUNTERING

> "Call it a character flaw – when under attack, I counter attack. Always." – *Richard Marcinko*

In tactically-correct fighting, there is only attack and counter-attack; there is no defense.

Defensive actions, whereby one protects themselves from attack but does not simultaneously return attack, are not tactically-correct, only acceptable. Certainly, if the moment doesn't permit you to offer an effective counter-strike, it is better to simply defend than to be struck, but the notion of *defensive* tactics is, at best, *defeatist* tactics because engagements are not bested without striking back (countering). Eventually, one can only defensively block and parry for so long before making an error, thus necessitating the need to counter-attack.

Countering simultaneously combines a defensive action with an offensive action while staying within reach or while moving out. Defensive counter-actions are classified in two ways: those that permit/assist an incoming attack to pass the imaginary central plane between adversaries (*parries*) and those that stop an incoming attack from reaching the imaginary central plane between adversaries (*stops*).

Offensive counter-strikes are classified in two ways: those delivered *directly* to the head/neck/body/thighs (trunk of tree) or *indirectly* to a hand/arm (branch of tree).

In this way, the complexities of close quarters chaos are reduced to the simple sophistication of executing a parry or block while simultaneously striking the most available/valuable target.

Countering Tactics

Back-ups avoid being struck by moving out of range while simultaneously attacking an opponent's attacking limb (indirectly).

Passing

Passing avoids being struck by parrying the incoming strike while simultaneously *indirectly* striking an opponent's limb, or *directly* striking their head/neck/body/thighs.

Stopping

Stopping avoids being struck by stopping (blocking) the incoming strike while simultaneously *indirectly* striking an adversary's limb or *directly* striking their head/neck/body/thighs.

Complex Passing and Stopping

Complex Passing and Stopping compound the defensive action with more than just a single offensive action. A *single pass/stop* strikes you only once during the execution of the counter. A *combination pass/stop* strikes you two times (using one action, though not two) during the execution of the counter. A *faking pass/stop* employs a defensive action coupled with a fake-and-strike offensive action whereby the fake is intended to draw out your adversary's defense and they circumvent it, preventing their ability to defend against it.

Counter-Grabbing

Counter-grabs are employed in conjunction with a pass or a stop, whereby the pass/stop action culminates with having seized your adversary's attacking limb, thus putting it under your control.

Counter-Trapping

The other side of a two-headed coin, counter-traps are executed the same as attacking-traps, whereby there is the intention to *trap* your adversary by forcing them to commit to a counter in which the next successive action on your part leaves them in a virtually-impossible position to effect a successful escape, defense, or counter.

DISARMING

> "Disarming a knife-wielding attacker is easy; the difficulty is in creating the conditions under which you can easily effect the disarm." – *Tom Sotis*

Contrary to the cooperative performances during martial arts demonstrations where the master easily, almost magically, disarms his attacker, actually disarming a committed knife-wielding attacker is the ultimate tactical challenge in personal combat. Indeed, disarming an opponent in real-time proves elusive because disarms are more dependent upon creating favorable conditions than on the action itself.

Lighting a fire with kindling and a lighter is easy to do. However, lighting a fire in a rainy windstorm requires you to find dry kindling and build a windbreak (create the conditions) so your lighter won't be extinguished.

The action of disarming (leveraging the blade from their grasp) is not difficult. Disarming a powerful and/or chaotic attack requires you to injure the attacker sufficiently (without exchanging strikes) such that you are able to seize control over their weapon-wielding hand. Once you have control and continue to tenderize them, the disarm is relatively easy to affect.

Sometimes people consistently fail at a relatively simple task and conclude it can't be done because they weren't able to create the favorable conditions required.

ACTIONS

PRINCIPLES

> "Principles and rules are intended to provide a thinking man with a frame of reference." – *Carl von Clausewitz*

Adaptation

> "Adaptability enforces creativity, and creativity is adaptability." – *Pearl Zhu*

Fighting is a spontaneous ever-changing event from moment-to-moment. *Tactical adaptation* means mentally and physically adjusting to the changing dynamics we are experiencing as the event unfolds from moment-to-moment.

When probing, we seek four response cues: overreaction, underreaction, delayed reaction, and failure to act. Similarly, fighters must develop the ability to recognize and appropriately adapt to the results of an action whether we are successful, unsuccessful, or fail.

Sometimes we are *successful with <u>no</u> immediate result* because our adversary's injury was insufficient to create an opportunity. This compels us to train with implements that cause no pain or injury so that our training opponent suffers no reduced capacity, and we must learn to remain in a mental state that is fully attentive to the moment.

Sometimes we are *successful <u>with</u> an immediate result* because our adversary's injury was sufficient to create a momentary advantage. This compels us to train with implements that cause pain (but not injury) so that our training opponent suffers momentary reduced capacity, and we must learn to recognize and capitalize on the opportunity while they are in recovery.

Sometimes we are *unsuccessful* because our adversary was able to mount a successful defense. This compels us to develop the ability to circumvent defenses and continue pressing the attack so as to force an adversary into an irreversible defensive state until they fail.

Sometimes we *fail* because we inadvertently revealed our intention and our adversary counters our action. This compels us to develop mental and physical recovery skills.

Centerline

"In knifing, centerline is your lifeline." – *Tom Sotis*

Centerline is the imaginary vertical plane that exists between you and your opponent. Guarding your centerline exercises the principle of minimizing your exposure to counter-attack.

This is most easily understood with a baseball metaphor: if you are on third base and wish to steal home plate, the best time to initiate your run is the very instant the ball leaves the catcher's hand because the ball must then travel all the way to the pitcher and then back to the catcher, giving you maximum time to reach home plate. But if the catcher tossed the ball up in the air and it quickly dropped back down for him to catch it, you would have virtually no time to reach home plate.

When you have attacked and your weapon reaches the point of intended impact, but misses (ball leaves the catcher's hand), the length of time it takes for your weapon to recover and be in position to protect your centerline is vital (ball back in catcher's hand).

Efficiency is measured not only in terms of delivery, but also in terms of recovery. Delivery is important because it is the means by which you dispatch your opponent, but recovery is of greater importance because it is how your opponent dispatches you.

Dynamics

"Dynamics define and measure the relevant forces at work."
– *Tom Sotis*

Pressure is a universal force in combat. The pressure from bombs explodes bodies apart. Mechanical pressure propels bullets and arrows, while human pressure forces knives into bodies, and delivers the impact blows from sticks or fists.

Gravity is a natural force which pulls at a constant, uniform rate. Pressure is also a natural force that can be either constant or changing.

Changing pressures create the dynamics of combat. A *combative rule of thumb* is that pressure is constant when the gap between fighters remains constant, and pressure increases as the gap contracts and decreases as it expands. Macro-dynamics refer to the general changes between fighters' bodies (think sumo wrestlers, or linemen in football crashing into each other). Micro-pressures refer to the contact and pressure between their limbs (think wing-chun kung-fu, boxing, or even arm-wrestling). Most fighters quickly learn that certain actions don't work against other actions. (i.e., certain blocks collapse against certain strikes). This occurs because techniques each have a ceiling of resistance to pressure and they break down when that limit is exceeded.

Students are often admonished by teachers that in order for techniques to work, they have to be executed "in the right context", but most students cannot do that because they aren't taught proper context to begin with. Rather than offering trainees admonishments and platitudes, or bad examples of context, I set about creating a way to measure pressures, and then subjected actions to the scale in order to establish the range of pressure(s) in which they are mechanically viable. Basically, I asked, "Which actions hold up under what kinds of pressure?"

The abstract ideas of context and dynamics become real and concrete when we replace them with the word "pressure". Different types of attacks are delivered with different pressures. The major factors that outline the dynamics are the *momentum* of the bodies, *angle* of delivery, *tension* of the action, the *timing* of the action, and the *intensity*.

VISION

> "Perception is merely reality filtered through the prism of your soul." – *Christopher A. Ray*

Our eyes play an enormous role in optimal human performance. Advances in neurology and technical equipment have produced corresponding advances in performance coaching.

During the eighteen months that I was afflicted with bi-lateral Bell's Palsy, I could not blink. This condition was less than ideal for everyday living, but it turned out to be *fantastic* when fighting. In a very short interval of time, I significantly improved my ability to detect what was coming at me because I wasn't losing vital moments of time to blinking. Soon, I began to maintain a single field of non-focus: my eyes didn't "jump-around" focusing on this, then that, then something else, etc.

Having learned to stabilize my vision was wonderful, and then ... I recovered from my illness. Soon I was blinking normally again. I had lost that which I had accidentally acquired, but not what I had discovered, so I set about attempting to re-acquire or invent it. I found that it was quite simple to avoid blinking by narrowing the opening to my eyes, thus lessening air contact with my eyeballs and extending the time between natural blinks to lubricate them.

Narrowing works well on objects at a bit of distance, and in the intervals between activating the momentary state of invincibility. In contrast, I sometimes purposely strive to activate an invincible state by consciously stretching my eyes wide open, creating a state of visual hyper-alertness that detects cues of movement with amazing clarity.

In addition to commanding control over blinking and saccades, your field of view is vital to survival during chaotic fighting. *Tactical wisdom* common in fighting methods advises that your overall field of view is 1) centered on or between adversaries, 2) the angle of your gaze tilts down sufficiently to see their feet (generally it is unnecessary to see over their head), and 3) when you stand up straight and look straight ahead, your field of vision extends out 180° to either side, straight up and down (imagine standing in a wide doorway between two rooms where you can only see the room in front of you, but nothing in the room behind you). As you bend forward and tilt your head down, the angle of your peripheral vision now tilts forward at the top and rearward at the bottom, permitting you to see slightly behind yourself at low angles and exposing people closing in on your flanks much sooner.

My approach to visual control was confirmed in 2007 when Dr. Joan Vickers published her ground-breaking book *Perception, Cognition, and Decision Training - The Quiet Eye in Action*. Upon reading about the Decisional Training Model, I discovered that AMOK! employed a remarkably similar approach.

MOVEMENT

> "There is no greater fame for a man than that which he wins with his footwork or the skill of his hands." – *Homer*

Technical movements are the specific physical actions that achieve tactical aims. While footwork is the primary technical means of tactical movement, it is *a* means, but not the sole means. All physical movements are technical in nature, and one might use the movement of the head, arms, or other body parts as well.

Technical Movement refers to overall changes of your positioning (moving from one place to another) and/or orientation (changing your angle of direction) achieved through actions such as pivoting, shuffling, stepping, skipping, sliding, twisting, jumping, and rolling.

Movement defines the positional relationship between you and your adversary. There is stillness when both are motionless, movement when only one person moves, and dynamic movement when both people are moving.

Attacking Movement - Stationary, Direct, Indirect

There are times when fighters hold their respective postures and inch ever-closer to being within range to strike each other without stepping, relying entirely on their mindset and reflexes. Because there is no demonstrative movement required to strike (each has already achieved their set-point), they are said to be attacking from a *stationary* platform.

When fighters are facing each other and an attacker advances toward their adversary *directly* in a linear fashion, they are attacking straight-on – moving along a straight line, the shortest distance between two points.

Other times an attacker may step off-line at a slight angle as they initiate their advance, aiming to complete the advance by making contact at an angle that is advantageous to the attacker (still facing his adversary's center straight-on) and disadvantageous to the responder (not facing straight-on). To attack *indirectly* is to seek advantage through misalignment of a fighter's center of balance.

Countering Movement - Avoid, Evade, or Intercept

A combative rule of thumb, for which there are always exceptions, is *not* to stand your ground when receiving an attack which benefits the enemy by being where he wants you to be at the moment he wants you to be there (X marks the spot). Holding your ground usually means dealing with 100% of your adversary's force. *Tactical wisdom* passed down by survivors is to *move off the X* whenever possible and still train for when you can't. In this way, you can employ maneuvers to lessen the force of an incoming attack as much as the situation allows to improve your positioning, but you are not dependent upon maneuvers to withstand the attack.

In moving to a better tactical position to counter, *Avoidance* means to move away so that your body is out of the adversary's reach, but you can still reach their extended hand/limb. This tactic offers the advantage of injuring your attacker with little chance of being injured in a mutual exchange. Even if you fail to strike their hand/limb, oftentimes you can still achieve a position of advantage from which to launch an attack while they are recovering their strike.

Evasion means to move off the spot you were on, but remain at a distance and where you can still both reach each other, either indirectly or directly to his hand/limb. As this maneuver results in both parties being able to reach each other, (the mutual-kill range of exchanging) one combatant will seek an advantage of positioning by moving off-line, preferable to the opponent's flank.

Interception means to advance directly into an attack. By advancing, one can often stifle an attack before it can gain sufficient momentum or position. It requires a steadfast intention to execute and when done so properly, can surprise the attacker and force them into a defensive dynamic. Interceptions can be executed indirectly (when the striking limb is in play) or directly to the attacker's torso.

POSTURE

> "It is often possible to decide the issue of a battle merely by making an unexpected shift of one's main weight." — *Erwin Rommel, The Rommel Papers*

Various forms of fighting assume a wide variety of postures depending on the form of the engagement. A Japanese swordsman wielding a katana stands and moves differently than a European swordsman wielding a longsword. A boxer stands and moves differently than a karateka, and so on.

The human body is a mechanical device that can adjust, prioritize, and emphasize its movements to accomplish a given task. Posture is the dynamic interplay between your feet/legs and your head/body and is governed by your balance.

Balance

Thanks to gravity, we can leverage ourselves off the ground to move; balance is the mechanism that rules over controlled movement. You are either in a state of balance (no pressure), off-balance (resisting pressure), regaining balance (recovery) or have no balance (falling).

To identify each of these states, try this exercise, but be gentle so that you can also observe. You should do each exercise in a safe area, in case you fall down.

- Stand with your arms relaxed by your sides so you are normally *balanced* (no pressure to resist except gravity)

- Do not move either foot while a friend pushes slowly into one of your shoulders. Your body will rotate away until you can't twist any further. If the friend stops pushing just when you are on the edge of falling but you do not fall, you are *off-balance* (resisting pressure).

- If they quickly release your shoulder, it will return to position to regain (recover) and if they push further past your off-balance point, either you will step to regain (recovery) or stumble and eventually regain (longer recovery).

- If you fail to regain your balance after stepping or stumbling, you will have *no balance* (fall to the ground).

Stance

Depending on what we wish to accomplish, the distance and angle between our feet either provides 100% of possible friction from which to leverage, or less, depending on how far out of ideal position for the task they are.

Besides the angle of my feet, the interplay between my heels and balls of my feet is also critical. When I want mobility, I move on the balls of my feet, and when I want stability, I'll be flat-footed. Our stance and footwork serve to support and mobilize our entire upper torso.

Spinal Alignment

Where the head goes, the body follows, so your head acts like a rudder that controls your posture and movement. The swivel in your back allows your head/shoulders to shift your balance, but you won't move off the spot on which you stand unless you shift your center of gravity off that spot.

BREATHING

"Breath is the link between mind and body." – *Dan Brulé*

Because breathing is something that we do without thought (unless there is a reason to think about it), breathing is automatic and natural – not something we have to learn. But it is something we can learn to do *better.* And through our mentors we discover that purposefully breathing in certain ways has a different effect or result on our external performance and/or internal experience which opens the doors to the depth and profound nature of breathwork. With regard to invincibility, I utilize two main approaches to breathwork: tactical breathing and circular breathing.

Amoks use stamina-building exercises for high-intensity combat-specific development by fighting full-out as they go through this progression:

- Full inhale and hold while fighting, then
- Full exhale and hold while fighting, then
- Fight during long slow constant inhale, then
- Fight during a long slow constant exhale, then
- Fight during 10-step inhale, then
- Fight during 10-step exhale, then
- Fight while hyperventilating until exhausted, then
- Fight until you can't go on, completely spent. Then,
- Give them the middle finger (can't break your spirit)

This and similar exercises build stamina so that we have more capacity and control under varied physically stressful conditions.

Back in the chapter on Engagement, we covered the elements that generate invincibility and those that leave us vulnerable. One of our vulnerable moments comes at the very end of expelling our breath. Because breathing and consciousness are connected, at the exact moment between exhale/inhale, it is as though your mind "blinks" for an instant (a crude metaphor). Just as we cannot afford to lose a millisecond through blinking at the wrong moment, neither can we afford to mentally blink by exhaling completely at the wrong moment.

To guard against this, Amoks practice *3/5th circular breathing*. This means we never fully exhale, only expending approximately 3/5th of our breath so that we always have something in reserve for surprise demands. In using the 3/5th rule, the fundamental breath we focus on converts the natural flow of our breath from in and out, or up and down, to the circular approach.

Riding a bicycle up and down hills illustrates this well. You know that if you stop at the bottom of a hill, it's going to take a lot of effort and lost time to get up the next hill, so you keep your momentum going as you coast through the bottom and start pedaling again as you're already starting up the next hill. Circular breathing occurs in the same fashion whereby: the "momentum of your exhale passes the bottom of your breath cycle and effortlessly begins your inhale that you then assist until the inhale passes over the crest and becomes the next exhale."

Focusing on a single point going around the circle can be a helpful visualization to "start the pump" and with every breath, new points appear on the circle until all points are connected and the entire circle is turning at once. Once your circular breathing is engaged, there is no longer any in/out or up/down breathing, just the constant cycling of respiration governed by the physical controls developed in tactical breathing.

I have the great fortune of being close friends with Dan Brulé, the world's foremost expert on breathwork and I strongly encourage everyone to attend one of his seminars or read his book Breath Mastery.

STRIKING

> "Weight, force and casual impulse, together with resistance, are the four external powers in which all the visible actions of mortals have their being and their end." – *Leonardo da Vinci*

The actions of striking are the physical manifestation of executing one's intention and as such, all strikes are born in one's mind and find their maturity in the resulting movement of one's body. The main elements of effective striking are non-telegraphic movement, speed, power, distance, timing, accuracy, and recovery.

When fighting, it is never a good idea to unintentionally reveal your intentions to your adversary. Hiding our intentions employs both deception (convince them you intend something else) and non-telegraphic movement (not letting them see what you are intending). When one's values are not in alignment with their strategy, internal doubt can keep the mind from "settling the platform of delivery" (akin to one short leg on an unstable table that constantly shifts its footing).

If one is unsure of the tactic(s) they have chosen to enact their strategy, even a tiny sliver of doubt can cause a minute hesitation at the moment of initiation, thereby causing a physical stutter in execution.

Even with values, strategy, and tactics firmly in alignment, at the moment one's intention galvanizes into action, the body must prepare itself (assume the posture and structure of launch) to deliver said action, which is aptly called the "set point".

Set points are an undeniable function that occurs when shifting commands between decision and action. Because set-points are impossible to prevent, it becomes necessary to learn to hide or disguise one's set point so as not to reveal one's intention before execution.

Finally, non-telegraphing striking demands seamless continuity between chambering (readying) a strike and delivering it. Any degree of "cock and fire" in the overall delivery exposes one's intention during the cocking phase.

Effective striking exercises the proper mechanics for the desired qualities of speed and power. One's own rhythm is a combination and synchronization of mechanical movement and breathing.

Effective striking also demands proper distance (not too far or too close), proper timing (not too early or too late), and accuracy (striking the intended target).

Effective striking entails not just the delivery of our strikes, but equally important, their recovery as well. Recovering from delivery means the time it takes in readying oneself to be guarding or striking again. Strikes have to be recovered after successfully finding their target in order to strike again. Strikes must be recovered when they are blocked or deflected, and when they miss making contact altogether.

TIMING

"Timing is everything." – *Marina Trent (to the author)*

The nature of fighting is a spontaneous, ever-changing event with randomly-changing dynamics to which one must constantly adapt. Rhythm is your body's own cadence. You express rhythm when you practice a martial arts form, or when you dance alone.

Timing is the application rhythm: like fighting an opponent, or dancing with a partner. Even if you have good rhythm, if your timing is off, you will experience a less desirable result.

Problems with timing arise because fighting is constantly changing and unpredictable, so it becomes advantageous to be able to naturally adjust to changes in tempo (the speed of rhythm).

RECOVERY

> "Toughness is measured by the speed at which someone recovers their state of readiness." – *Tom Sotis*

What is toughness? Sometimes when you punch someone, they stagger back or cover up for a moment or two. Other times, some people just look back at you with barely a hint that you even struck them. I think we can agree the second type are tougher, and our yardstick of that measure is in their recovery time.

There are seven types of recovery that lend to peak performance and they are: tactical, visual, mental, physical, emotional, breathing, and lastly (but of vital importance), centerline.

- We develop *tactical recovery* by practicing when and how to disengage from chaos or a serious mistake.
- We develop *visual recovery* by practicing to re-establish our visual field and processing movement.
- We develop *mental recovery* by learning to regain mental stability and re-establish our intention after a distraction or following an action.
- We develop *physical recovery* through exercises that cause us to make mistakes so that we can learn to re-establish our balance and posture.
- We develop *emotional recovery* by practicing to quell our responses to our emotional triggers. We have emotional stability (control) when others are unable to tell how we feel by our expression and demeanor (poker face).
- We develop *breath recovery* by practicing to re-establish circular breathing if we are forced to completely expel our breath.
- We develop *centerline recovery* by practicing exercises that challenge our practice partners to exploit our centerline, thus forcing us to be painfully aware when we fail to do so.

PART III: APPLICATION

A DISRUPTER

> "Everyone is born with an instinct of success, and ability to make it, but only the ones who think out of the box succeed." — *Unarine Ramaru*

The martial arts / combatives industry is unregulated insofar as anyone can invent their own system, open a school, charge for classes, print certificates, award ranks and titles, and build a business in whatever way they see fit. Some do this with integrity and excellence...though they are indeed the exception, not the rule.

Martial standards vary and are often inconsistent within a school, between schools of the same system, and between organizations. In other words, what constitutes a black belt varies far and wide. Rank and title hold value in terms of status/prestige for those who deem such recognition important, so conflicts arise when their egos collide. Disputes over how someone was awarded whatever rank or title and by whom are the foothold of all martial politics, the most disruptive and divisive element within the discipline.

If martial politics over rank isn't bad enough, situations become toxic when compounded by money which turns the buying/selling of ranks that are supposed to be skill-based into income-based promotions. This explains how a thirty-year-old grandmaster gets his rank while a fifty-year-old, who trained for thirty years, might be of a lesser rank.

Combining the power of authority (rank) with money creates the footholds of corruption and martial prostitution. Some owners of martial franchises intimidate or bully the school owners into agreeing or complying with unfavorable business policies. Many martial arts school owners sell programs designed not in the best interest of their students, but to profit at their expense ... and ultimately, peril.

Fortunately, there are still some martial arts instructors who train diligently, teach openly with integrity, and exemplify the values they espouse. I hold them in the highest regard. Indeed, they may be hard to find, but they are worth every bit of the search.

> "A vision without a strategy remains an illusion." – *Lee Bolman*

I built AMOK! and avoided the systemic flaws that plague martial organizations by creating a proper combative training business. I accomplished this by:

- Inverting the organizational structure from an enclosed pyramid to a funnel that provides open-ended development and support
- Reversing the order of training activities and developing a functional method of integrating skills
- Abandoning the awarding of extrinsic ranks and titles, and offering only the intrinsic rewards of gaining skill and fostering brotherhood
- Establishing licensing requirements to meet professional standards, policies to establish fair practices, and making them 100% transparent
- Instituting a system whereby AMOK! trainers all pay a modest monthly flat-fee with no percentage of their earned income so that finances never become an issue
- Employing the opposite means of delivery from brick-and-mortar fixed locations to mobile delivery in convenient venues

Inverted Structure

> "Not knowing how to proceed, I simply attempted the opposite of what has repeatedly failed." – *Tom Sotis*

In forming an organization, I wanted to avoid the inevitable pitfalls most commonly plaguing many martial arts systems and their organizations. The three main elements leading to organizational disruption/destruction are martial organizational structures, martial ranks/titles, and non-professional (sometimes unprofessional) business practices and standards.

Most martial arts organizations employ closed-containment pyramidal structures with students at the base, instructors up one level, masters above them, and a grandmaster at the top. Typically, the most powerful personalities end up jockeying for position in a limited space, inevitably causing friction (politics), which eventually becomes untenable. The system fractures and one or more of the Masters split from the organization to form his own, taking a number of members with them. If this doesn't occur when the Grandmaster is alive, it is almost certain to occur immediately following his death. Sadly, few organizations seem to escape this fate.

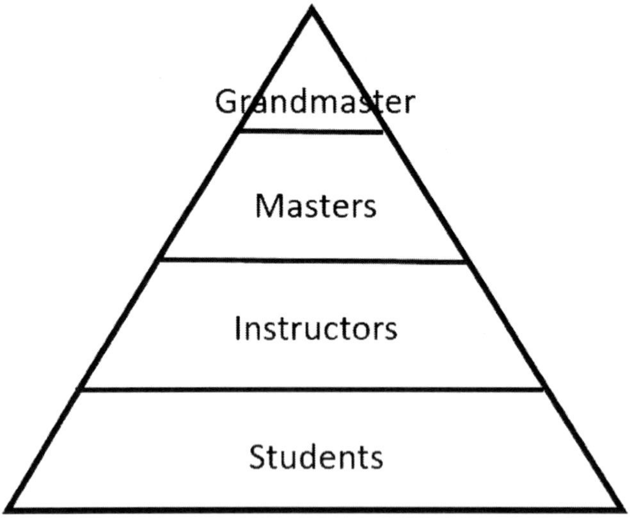

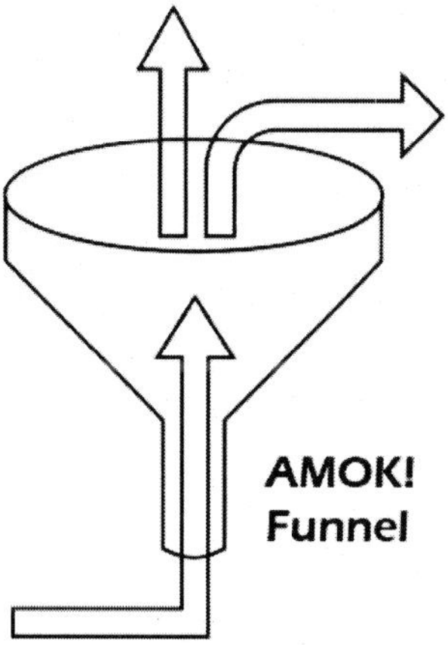

I did the opposite when creating AMOK! by inverting the pyramid and employing an open-support funnel-shaped structure where my job, and that of the trainers, is to get people up through the stem of the funnel (functional with core skills) and then help them to develop up in the cone, in whatever direction suits them. There is no limit to their development and they can jump out of the cone to start their own organization. This way, the people we develop and support don't have to tear down our house to do their own thing.

Reverse Method

"Sacred cows make the best hamburgers" – *Mark Twain*

Imagine if a soccer coach had his team doing lots of basketball footwork drills, practicing some football plays, and then playing a little soccer at the end of each session. As you might think, when the soccer team plays against other teams, their performance will be deficient, much lower than if they had proper coaching. Still, some naturally-talented athletes would become great players in spite of their training, and the average player would be condemned to mediocrity no matter how much work he or she put into the task.

As hard as it is to fathom, with few exceptions, the skill-development methods of martial arts systems employ this broken model; they have inexplicably failed to adopt the advances of modern sports science. To be fair, traditional methods should not make even the slightest changes because their task is to preserve the art, broken though it may be. But the vast majority of non-traditional methods that claim to be self-defense worthy are still using antiquated, dysfunctional methods of skill integration.

The three main elements of methodology are the *order, emphasis, and relationships* between the three types of activities of *training, practicing, and sparring.* These elements are supposed to have a mutually-supporting components, thereby creating a synergistic effect that is often plagued by *functional disconnects,* reducing performance when the relationship between activities is incongruent (like gears that don't mesh).

Martial Methodologies

These methodologies begin with training exercises to develop movements of one's body that mimic how they desire to move in fighting, oftentimes emphasizing style, and inhibiting body mechanics over functional form.

They proceed to practice drills that barely resemble fighting – with static cooperative attackers or patterned activities – unlike actual fighting, with dynamic uncooperative attackers who don't move in patterns. Sometimes their practices use moves that students have not yet learned in training.

Finally, they finish with some form of controlled sparring in which the inhibiting training movements and unrealistic practice moves are supposed to produce improved fighting functionality.

Not only are the activities themselves deficient, but the relationships between them are divorced! It is impossible not to notice how the training moves don't align with the practice drills, which in turn, don't align with the fighting activity. This is true of some more than others, but all the empty hands and weapons methods that I studied all suffered from these systemic defects in their methods of integration.

My questions have always arisen out of genuine curiosity and the desire to understand what the teacher wanted us to learn, though in many cases, I was dismissed, chastised, ignored, ostracized, held back, punished, or even asked not to return for asking questions about systemic defects in class.

"Relentlessly prune bullshit." – *Tom Sotis*

Instructors of those methods are well-schooled in double-talk explanations that students either find acceptable or don't challenge, and some instructors themselves blindly believe what they profess.

I attribute this to a form of martial bias, whereby people don't vigorously question what they are learning, and indeed, many are admonished *not* to question the system because they "haven't mastered it yet", so it can't be explained to someone at their level ... and other such deflective and condescending hogwash. Some martial arts methods are little more than a haphazard collection of gears thrown into a box.

"If there's a way to do it better...find it." – *Thomas Edison*

Functional Methodology

Frustrated with dysfunctional methods, I abandoned all of them and set upon finding a better way. I told my guys we would shelve all training exercises and practice drills in favor of only fighting for an indeterminate amount of time (we just played soccer).

Within six months, my group had dwindled from forty guys to about a dozen. During that stage of attrition, we learned well how to attack each other (learned to "score"). Soon, we began to discover and enact counters to those attacks (defense plays). Soon after that, we discovered how to circumvent counters (offensive plays), and *by fighting it out*, eventually, the rock-paper-scissors of tactics was discovered. The corresponding training exercises to improve our physical capacity became obvious (soccer footwork). AMOK! as a fighting method was reverse-engineered in its natural environment.

Completely the reverse of martial methodologies, AMOK!'s ultra-functional fighting method begins with *experiential fighting* to expose tactical weaknesses, then *experimental sparring* to discover (not invent) solutions, which then dictates authentic *skill-developing practices,* and the corresponding *attribute-development training.* Because AMOK! practices and training activities are done in sparring mode, they are *functionally-authentic,* unlike the false dynamics of staged practice drills common in martial methods.

Furthermore, because AMOK! employs a fight-based platform for virtually all activity, it *permits no functional disconnects* between the main activities because the underlying activity is continuous throughout, whether you are fighting at 100% intensity, sparring at 90% intensity, still sparring while practicing specific moves at 70-90% intensity, or still sparring while training the elements of distance, timing, and mechanics (75-100% intensity).

While I did the work to develop AMOK!'s unique and ultra-functional methodology, I credit David Gould for teaching me the principles of tactical exploration so that I could follow my own path to combative functionality.

The AMOK! methodology was developed independently of the ground-breaking research for peak performance under pressure done by Dr. Joan Vickers and outlined in her book, *Perception, Cognition, and Decisional Training.* I am quite pleased my work was validated when I observed the parallel "top-down tactical approach" between AMOK!'s method and Vickers' Decisional Training Model.

Integration

"Don't imitate or accumulate, but integrate!" – *Bruce Lee*

AMOK! is a response-based approach to integrating functional skills which demands that activities possess three principal ingredients: authenticity, context, and deliberateness. *Authenticity* means establishing the desired/required state of mind and dynamics. *Context* is that which gives meaning and significance, perhaps even defining the very nature of a thing itself. *Deliberateness* describes an action that possesses a trait of thoughtfulness in executing a purposeful intention.

Authenticity

"You must get wet to swim." – *Unknown*

Imagine you must learn to play water polo in a raging river. How much would you expect to improve your performance if you spent 90% of your practice time on a volleyball court helping each other volley and setting up opportunities for each other to spike the ball? Once in the raging river of the water polo game, you'd be so busy trying to keep your head above water that it would be very hard to pull off volleyball techniques. First you have to jump in and keep your head above water, *then* you can learn to play while you are *in* the water. Sure, you might find a spot in the river where the water slows down to work on something, but you stay in the water because no matter what you are working on, swimming is not like standing. The raging river is a metaphor for the opposing will of your adversary, and the unpredictable nature of fighting.

Authenticity is not only about establishing the natural, physical environment of spontaneous fighting; equally vital is establishing the natural mental environment too. *Authenticity* means establishing the desired/required state of mind. When fighting, you must remain 100% attentive and adaptive to the spontaneous changes of the event. If you face off with an adversary and you look away, even for an instant, you are likely to perish in the very next moment. You pay *attention* to your adversary and this is the state of awareness that needs to be activated and trained. That is why virtually all of AMOK! activities are fight-based because the authenticity of the situation automatically sets the right frame of mind.

Context

"I think it's important for people to always understand what is the context is in what is being said, because that obviously determines what folks are talking about." – *Roland Martin*

The Mystery Technique Puzzle

You have not seen the picture, you don't know the shape of the puzzle, and you don't have access to the pieces. You get just one piece at a time and you work with it until you find where it fits. Sometimes it won't fit until later and sometimes it will never fit anywhere, but still you try to find where it fits until you put it aside. Even pieces that will fit reveal no clue as to where they go because you have not seen the whole picture. For example: you get a red piece – do you know where it goes? No, because you haven't seen the picture yet. *Individual pieces lose their value when not in context with the whole of which they are part.* After an exhaustive effort, the picture begins to take shape and partially reveal itself. Eventually, you complete enough of the puzzle and solve the mystery. It turns out to be a farm scene with green grass, blue skies, brown cows, and a red barn. Sometimes after all that work, you don't like the picture. This is what it is like to learn much of the martial arts.

> *Trainee: Is this a good technique?*
>
> *Me: If you can show me a time and place where that is a functional solution, then yes, and outside of that, probably not.*

The Functional Tactical Puzzle

You are asked to choose a picture that you like. You choose a farm scene with green grass, blue skies, brown cows, and a red barn. Put all pieces face up on the table. Select the edge pieces and assemble the frame. Group the colors according to the picture. Assemble the pieces. If you drop a piece on the floor and you pick it up and it is red, you know roughly where it goes and where it doesn't. *Individual pieces gain their value when put in context with the whole of which they are part.* This is what it is like to learn AMOK!

> "The important thing about any word is how you understand it."– *Publilius Syrus*

Tactical Context

Even effective techniques, if learned out of context, will likely be used the wrong way or at the wrong time, and fail with potentially disastrous consequences. Equally so, drills learned out of context promote flawed expectations and develop a false sense of skill.

The vast majority of "experts" teaching self-protection are unable to define, and then compare and contrast the three interrelated levels of strategy, tactics, and techniques.

- Strategy - the overall objective, *(your goal)*
- Tactics - the plan to achieve the objective *(your intention)*
- Technique - the spontaneous assembly and application of movement *(your action)*

The vast majority of "experts" teaching self-protection are unable to define, and then compare and contrast the three main activities of opposition, practice, and training.

- Opposition - opposing wills to minimize mistakes *(fighting and sparring)*
- Practice - replicating tactics to gain tactical skill *(paired drilling)*
- Training - exercises to improve physical attributes *(paired and solo exercising*)

The word *technique* is perhaps one of the most used words in the martial arts profession, so I conducted a Facebook experiment asking for accomplished martial artists to provide its definition. Over 100 people responded and *not one* of them got it right. I challenge you to conduct the same experiment and you will find that my definition always holds, while anything different doesn't hit the mark as accurately.

If the experts lack a grasp of the fundamental framework of fighting and the elements comprising tactical context, how then are their students to learn it? Accidentally, incidentally, arbitrarily, or never?

Deliberateness

> "There is no magic. There is only knowledge, more or less hidden " – *Gene Wolfe*

There are no supernatural powers, incantations, spells, or any other pseudoscientific practices that develop skill. Skill development lies entirely and naturally in one's own mind and body, and this is evident in how babies learn to walk long before they can understand or develop beliefs in man-made constructs.

If there is anything close to magic – and I don't mean something supernatural, but something extraordinary – it is understanding the key to *the biology of skill development.*

When a baby is learning to walk, he or she very *deliberately intends* to take a step, sending a stream of impulses down the axons (nerve fibers) to actuate movement in the muscles. In the early stages of development, some signals "jump off" the axons (which is why other muscles twitch) resulting in fewer and/or delayed signals.

And this is how the magic of mislabeled and elusive "muscle-memory" occurs. Every time you take a *deliberate* action, the body produces a substance called myelin that begins to coat the axons, acting like insulation and improving signal transmission. With every deliberate step, myelination increases until the *intention/action* are achieved "as one", producing an action that springs forth from your will, seemingly bypassing conscious effort, and resulting in "perfectly natural movement" executed without the hindrance of processing.

Most people believe that repetition increases skill development. Initially, this is true, but then all of us experience plateaus in performance of one kind or another. Plateaus are the ceilings of our performance whereby even with exhaustive repetitive effort, gains are negligible, if there are any at all. Then why do skill plateaus occur if you are still doing the repetitions, or even when doing many more? Because...

> "Only deliberate repetitions develop skill; the rest just burn calories." – *Tom Sotis*

Deliberateness describes an action that possesses a trait of thoughtfulness in executing a purposeful intention.

This following example of popping soap bubbles illustrates the difference between repetitive and deliberate actions.

Imagine you had a machine that would blow soap bubbles out in a straight line, one bubble each second, and your task was to pop them with a pointed object like a pen. You could position yourself, hold up the pen, and once you got the timing down just set yourself on autopilot and pop, pop, pop, pop, pop, pop, pop for as long as you like, and while indeed you may perform hundreds, thousands, even tens of thousands of repetitions. In the end, you might have popped a lot of bubbles, but you didn't become a more skillful bubble-popper.

If, however, I held the bubble wand and blew out a bunch of bubbles that flew out in a random order, then you would have to *select each one, "eye it out" (recognize spatial movement) and deliberately pop it* before it hit the ground. In this fashion, you would become a more skillful bubble-popper, able to adjust to randomness.

Opposite Delivery

> "If everybody is doing it one way, there's a good chance you can find your niche by going exactly in the opposite direction." – *Sam Walton*

In the early '90s, I relocated back to my home state with all that I owned – some books and my weapons – and I got a job as a doorman at a popular nightspot on the waterfront. I knew I wanted to be teaching, but I had no resources to start a school.

Martial arts attract a small percentage of clients from the general population, firearms attract an even smaller slice, and knife methods attract the least. Martial arts schools tend to exist in fixed locations (retail space) with ongoing programs in competition with other methods in their geographic area. Firearms training tends to exist in fixed locations (shooting range) with ongoing programs in competition with other ranges/instruction in their geographic area. Both martial arts schools and shooting ranges require a substantial investment, monthly overhead, insurance, and maintenance.

So, I became a Personal Martial Arts Trainer. I made business cards, fliers, and brochures and drove around, putting them up at gyms, colleges, bulletin boards, coffee shops, etc. Instead of operating from a fixed location, I decided to be mobile and deliver one-off or short-term courses directly to my clients by driving to train them anywhere convenient for them.

At the time, for $50 I would drive to anyone's home, office, gym, or wherever they wanted and give them a one-hour private lesson. Two guys who owned a gym started to train with me, and then a couple of guys from their gym started (as of this writing Glenn Chavious still trains with me, as do his sons). Between a little legwork and word of mouth, within three months I had twenty clients. On just twenty hours of "work" (dare I even call it that) at $50 each ... Do the math. I got to be mobile, I made my own schedule, my clients were happy, and I was earning more than I could be making at any job for which I was qualified at the time. This worked out great during the week, but I found most private clients didn't want to train on the weekends. I wanted to maximize my income, so I had to develop a program that was weekend-oriented.

Getting private clients with advertising was like fishing with a net and I was doing reasonably well considering I had a lousy net (amateur marketing) but filling seminars is a completely different ballpark than getting individual clients.

Rather than trying to attract hosts by casting a wide net of advertising, I realized I could be spearfishing instead. So, I formed strategic partnerships with martial arts school owners to host my seminars. I contacted them directly and set up a meeting for coffee. Because I did not have a school, I was not their competition, and because my courses complemented theirs without conflicting, they never had to be concerned about losing students to me. This worked out extremely well. Soon, my calendar was filled with weekend seminars and I was traveling to other countries.

My work with government agencies, military groups, and law enforcement agencies increased in frequency, as did my work among tactical shooters (not competitive shooters). As I began developing dedicated knifers, I instituted a professional licensing program, and the mobile-instructor model has proven to be a successful alternative to operating a brick-and-mortar location. To date, I have taught over 1200 seminars in twenty-five countries, far more than any other knife instructor in the world, and AMOK! is the world's leading edged weapons training company.

ABOUT THE AUTHOR

Tom Sotis has been training in various fighting methods since 1970, and is now recognized as the leading edged weapons instructor in the world. Having taught over 1200 seminars in twenty-five countries, Tom is this generation's most prolific instructor of knife combat, earning world-wide acclaim for his unique theories and coaching methods.

In the early 1980s, Tom trained as a criminal investigator in Los Angeles, California, and then returned to Rhode Island where he owned and operated Metro Criminal Investigations for ten years, working mainly in the areas of asset recovery, personal protection, employee integrity, internal investigations, undercover operations, fugitive recovery, and homicides.

In 1994, Tom was the first American private contractor to train the famed Spetsnaz Counter-Terrorism teams, as well as personnel from the Ministry of Internal Affairs, Russian Criminal Investigation Division Homicide & Major Crimes, and numerous other units. Tom was recognized as an honorary member of the famed Red Berets, appointed Special Advisor to the Ministry, and honorary Chief Instructor of the CID/HMC division.

In the Philippines, Tom was the first non-Filipino adopted into the Dogeaters warrior tribe and to have his knife system recognized by the Bacolod Negros Arnis-Kali Escrima Federation. With dedicated training and study, Tom also earned master ratings in five traditional Filipino martial arts.

Internationally, Tom's work experience includes, but is not limited to, Cambodian Special Forces, Mexican Prison Guards and Riot Control, Greek Coast Guard, New Zealand Prison Guards, Norwegian Law Enforcement, South African Military and Police Forces, and Spanish Law Enforcement.

For over twenty-five years, Tom has worked with numerous US government agencies, specialist military teams, and various levels of law enforcement agencies. This list includes: US Intelligence Agencies, US Special Forces, US Secret Service, Federal Bureau of Investigation, Pinellas County Sheriffs, Massachusetts State Police, numerous municipal police departments, Corrections Officers and Special Response Teams.

Tom has appeared on Russian National Television, and on radio talk shows in South Africa, USA, and New Zealand. He has been featured in the books *Masters of the Blade, Filipino Martial Arts Masters and Grandmasters*, and *Just Breathe*. Tom has been featured in *Blitz, Boxing, FHM, Ring*, and *SWAT* magazines.

In addition to his extensive martial and combatives background, Tom has also cross-trained with various firearms, in knife-only survival and bushcraft, as a locksmith and vault technician, in scuba nitrox and rescue diver, and sailing.

Concentrating in a particular area of psychology, Tom became a certified motivation analyst licensed to administer and interpret the Reiss Motivation Profile®, the world's first scientifically-validated and most accurate method of personality profiling and predicting behavior.

SPEAKING ENGAGEMENTS

Tom is an engaging speaker with a unique background of compelling true-life experiences. He shares valuable life lessons learned from his adventures while traveling, studying, teaching, and fighting in some of the world's most dangerous places. When speaking, Tom brings listeners into his experience with vivid recollections and thoughtful reflections that provide insight into the mind of one of the world's most lethal men.

Tom openly shares life lessons learned from bounty hunting, and street fighting in Los Angeles, CA, and hard lessons from working undercover in Mexico. He has the distinction of being one of the very first Americans to travel deep behind the "Iron Curtain" in Russia where he was challenged by the famed Russian Spetsnaz

Special Forces. Another popular topic, Tom relates the stories and lessons from his travels to the Philippines where he studied several systems of knife fighting, and became the first American to have his knife system recognized by the old masters.

Tom recounts adventures from teaching in South Africa where he went on patrol with high-speed reaction units, and from other journeys, such as facing challengers high in the majestic mountains near Mt. Everest in Nepal. Listeners are usually surprised to learn that, even in the USA, on rare occasions, duels with real weapons still occur, and Tom shares the details of fighting in one of these duels.

Tom captivates listeners with his energy and passion. He speaks from his heart with unflinching honesty, sharing his fears and triumphs through the remarkable experiences of his life. Tom is one of the most stimulating and authentic speakers available with fascinating stories that listeners are sure to enjoy.

BUSINESS COACHING

Tom offers executive coaching that combines the powerful insights of the Reiss Business Motivation Profile® with his Model of Invincibility – a framework for perfect performance in knife combat adapted to a business model that explores the relationships between, and provides visibility in the important areas of Company Values, Leadership Strategies, Management Tactics, and Service Operations.

As the keys to invincibility are universal, Tom trains executives to optimize their personal performance. Whether it is for an interview, presentation, testimony, or any other reason that you want to maximize your personal performance, *Invincible* teaches you how to be your very best when you most need to be. When making decisions under the highest degrees of stress, Tom will share with you the method he developed, one that is used by special operatives and professional fighters to focus and stabilize their minds when the situation turns chaotic.

Invincible offers a powerful and unique team building experience because engaging in this activity requires trust above all other things. Knife training (with safety trainers, of course) is "the act of personal bonding in which physical skill is the byproduct".

SPORTS COACHING

Designed for both athletes and coaches, Invincible sports coaching offers an unparalleled platform of high-performance athletic coaching that brings together two of the world's most powerful methods of personal development. By combining the powerful insights of the Reiss Sports Motivation Profile® and *Invincible*, Tom delivers insight into our deepest psychological drivers combined with unique physical practices for superior performance under the highest types of stress.

Never before have these two elements of mental and physical performance been combined to develop human potential to such an exceptional degree. Tom's method of physical culture, presently utilized within the special operations and intelligence communities, incorporates the neurological approach of Decisional Training utilizing visual and attention control with additional components particular to prevailing in life-or-death combat. These exceptional qualities make Tom's program the ideal choice for top athletes and special operatives seeking to maximize their potential.

Going beyond Perception-Cognition elements addressed in Decisional Training methods, The Matrix of Invincibility maps the functioning elements of the heart (values), the mind (mentality), and the body (movement) to identify and repair "functional disconnects" that cause reduced mental performance. *Invincible* coaching employs a more diverse model of activities in the Cycle of Physical Optimization for peak performance, and further employs a parallel framework: the Cycle of Mental Optimization, to achieve perfect performance more frequently, and for longer durations with regard to any athletic pursuit.

REFERENCES

Morgan, Forrest E. (1992). Living the Martial Way: Manual for the Way a Modern Warrior Should Think. Barricade Books.

Reiss, Steven. (2000) Who am I?: 16 Basic Desires that Motivate Our Actions Define Our Personalities . Penguin Publishing Group.

Vickers, Joan (2007) Perception Cognition and Decision Training

Brulé, Dan (2017) Just Breathe. Simon & Schuster, Inc.

Sapolsky, Robert, Ph.D. (2017) The biology of our best and worst selves. TED https://www.ted.com/talks/robert_sapolsky_the_biology_of_our_best_and_worst_selves

ACKNOWLEDGEMENTS

I am thankful for my mother, Marina Trent (RIP) who gave me life and taught me to stand both firmly and compassionately, on my own two feet; and for my father, Charles Sotis, who did his best.

I never imagined I would be so lucky to have such a wonderful and loving wife as Lucia, my partner in life who has always stood by me, for without her, I would be lost, and for my exceptional son, Jake, of whom I am most proud and cherish every day we have together.

With regard to AMOK! I have a very long list of people I will mention when I write a book about AMOK! That said, I am compelled to recognize Dan Neilan from the USA, Randall Hodges in Thailand, and Mattias Persson in Sweden, as the three men who sacrificed the most and worked the hardest with me to put AMOK! on the map, and to Tim Gillett for hosting my very first knife seminar.

I have been fortunate to have had many fine teachers, and among them, I wish to offer my respect and appreciation to Eugene Coutu (Kenpo) who taught me about the importance of integrity, to Frank Masiello (Kuntao) who taught me about resisting martial pseudo-science, to P. Greg Alland (Sinatirsiawali) who embodied and inspired the quality of generosity, and to Nene Tortal (Dekiti Tirsia Siradas) who taught me the virtue of humility.

I feel martially-privileged to have trained with David Gould (Lameco Escrima), a consummate warrior of exceptional skill for whom I have the highest respect, who shared with me the principles and tools to achieve ultra-functionality without restricting my independent path of exploration and development. I can never pay back the gifts I received from Guro Dave so, following his example, I pay it forward.

I am thankful for my mentors who each, in their own unique way, made a significant impact on my life. My close friend, John Ebert (RIP), taught me through firewalking not to accept the illusion of absolutes. My good friend, Dan Brulé, taught me through breathing and by his example, how courage inspires a life of adventure. My AMOK! brother, Jeroen Terpstra Ph.D. and I have studied with each other for fifteen years. He introduced me to Reiss Profiling and helped me to better understand human behavior, especially with regard to violence. I was fortunate to meet and work with Scott Sonnon in person for only one weekend (so far), but his pioneering book on the 3-Dimensional Performance Pyramid has been the conceptual basis for a great deal of my work.

I am grateful for the time I've spent with Tony Nester, the renowned survival instructor. Tony and I trained together and became close friends while sharing our respective survival expertise from which I enjoy a powerful sense of independence and empowerment.

I am thankful for my AMOK! Brother, Andrew Smith from Australia, who protected AMOK! during the hostile takeover attempt and instituted the licensing procedure to meet professional standards. Without Andrew, AMOK! might well have suffered irreparably, and instead, gained even firmer footing. I am also very thankful to my AMOK! Brother, Glenn Chavious from the USA, for his business guidance, and support. Ultimately, he convinced me to write this book.

My deepest gratitude to the hosts, and all you fine folks around the globe who have attended my seminars. I recognize there are great demands on people's time and finances, and when you choose to spend both of those valuable personal resources to attend my seminar, I endeavor to give you 100% of myself, my very best, every time. I thank each and every one of you all for populating the events that shaped the course of my life. I love you for it.

GLOSSARY

Actions – the physical actions taken to achieve a tactical aim

AMOK! – the trademarked name of the combative method founded by Tom Sotis

Amoks – people who train in AMOK!

Arnis – generic term for some Filipino martial arts

Authenticity - replicating conditions to sufficiently establish the desired/required state of mind

Combative rule of thumb – a broadly accurate guide or principle, based on anatomy, experience, or practice rather than theory.

Context - the circumstances that form the setting for an event, statement, or idea, and the frame of reference within which that event, statement, or idea can be fully understood and assessed.

Deliberateness – describes an action that possesses a trait of thoughtfulness in executing a purposeful intention; a myelin-building activity.

Deception - the manipulation of appearances such that they convey a false reality

Dogeaters – indigenous Filipino warrior tribe

Engagement - encompasses the moments between initial physical contact and last physical contact

Escrima - generic term for traditional Filipino martial arts

Exceptions to rule of thumb – exceptions to a broadly accurate guide or principle, based on anatomy, experience or practice rather than theory.

Guro – generic term for Instructor in varied Filipino martial arts

Jeet Kune Do - "The way of the intercepting fist" is the name of Bruce Lee's martial art

Kali - generic term for some traditional Filipino martial arts

Myelination - the process of forming a myelin sheath around a nerve to allow nerve impulses to move more quickly

Practice – activities to gain skill in applying effective techniques

Protocols – established courses of action designed to achieve best possible outcomes

Preparations – obtain the items and training sufficient to accomplish a given task

Pre-engagement – encompassing the moments between initial communication and engagement

Post-engagement - encompassing the moments following engagement

Readiness – the state of being fully prepared for something

Sparring – activities to minimize mistakes and improve one's functionality against opponents

Strategy – an objective designed to achieve a major or overall aim

Tactics – plan to achieve the strategy (in general), momentary intention (personal combat)

Tagalog – indigenous Filipino dialect

Technique – the spontaneous assembly and application of movement

Training – activities to improve one's physical capacity and attributes

Tuhon - generic term denoting "chief instructor" of Filipino martial art

Made in United States
North Haven, CT
10 November 2026

22102491R00104